ACTIVATING OUR
12-STRANDED DNA

"Ruslana's masterful integration of sacred geometry, gematria, breathwork, DNA research, and sound creates a treatise of unparalleled excellence. This is the kind of book where each revelation draws you deeper, each chapter beckons you to uncover the next secret. And just when you think you've grasped it all, you'll find yourself compelled to revisit each chapter, uncovering new layers of wisdom."

MAUREEN J. ST. GERMAIN, AUTHOR OF
LIVING YOUR BEST 5D LIFE

"This beautifully written, highly engaging work effortlessly synthesizes biological science, healing, and consciousness to empathetically provide a 'blueprint' for both spiritual and personal development. Highly recommended for anyone interested in the larger horizon of spiritual evolution, global healing, consciousness, and their underlying foundations in biology."

PETER MARK ADAMS, AUTHOR OF
THE POWER OF THE HEALING FIELD

"*Activating Our 12-Stranded DNA* is a journey of exploration into the elusive space between the mind and beyond the mind. Remennikova's deeply informed writing allows archetypal memories, wisdom, and intuition that we thought we did not have to be brought into our present consciousness. This enables us to recognize patterns in our personal lives that mirror and replicate much larger patterns that span all times and universes."

ROBERT HENDERSON, AUTHOR OF
EMOTION AND HEALING IN THE ENERGY BODY

"The brilliant Ruslana Remennikova lasered her encyclopedic knowledge of the connections between our DNA and the cosmos in her new book. Within your cells, chromosomes with their noncoding DNA hold transformative potential when bathed in high-caliber frequencies, especially optimal sounds. Our life-giving friend—water—remembers, has intelligence, and couples with DNA to enter the fourth dimension. Rich in ritual, studded with illustrative autobiography, Remennikova creates a bridge for our current yearning to merge science with the mystical."

BERNARD BEITMAN, M.D., AUTHOR OF
MEANINGFUL COINCIDENCES

"Ruslana masterfully combines science and spirituality, showing how our DNA is interconnected with almost everything in the universe. Her engaging and authentic stories simplify complex ideas, making you nod and smile as you read them. Her rich book calls for spiritual evolution, encouraging you to activate your genetic potential through frequency medicine, visualizations, and exercises."

DAVID BARRETO, AUTHOR OF
KARMA AND REINCARNATION IN THE ANIMAL KINGDOM

"We are entering a new age of scientific discovery in which we are beginning to learn that physics and biology necessarily point to deeper truths that are revealed when we see these disciplines in their true light. These deeper truths were apprehended by ancient teachers like Pythagoras, and now Ruslana Remennikova is applying them to genetic science in a way that will inaugurate a revolution in our understanding of biological life."

ARAM ALCALAY, AUTHOR OF *GRAVITATION*

"As an astrologer who knows that DNA is primary and that this author has so many of the keys (music, vibration, number, science, animals, health, history, myth, and love), I look forward to reading this book again and again. Like the discovery of a new star, I can feel the possibilities gathering."

FREDERICK HAMILTON BAKER, AUTHOR OF
ALCHEMICAL TANTRIC ASTROLOGY

ACTIVATING OUR
12-STRANDED DNA

Secrets of
Dodecahedral DNA
for Completing
Our Human Evolution

A Sacred Planet Book

Ruslana Remennikova

Park Street Press
Rochester, Vermont

Park Street Press
One Park Street
Rochester, Vermont 05767
www.ParkStPress.com

Text stock is SFI certified

Park Street Press is a division of Inner Traditions International

Sacred Planet Books are curated by Richard Grossinger, Inner Traditions editorial board member and cofounder and former publisher of North Atlantic Books. The Sacred Planet collection, published under the umbrella of the Inner Traditions family of imprints, includes works on the themes of consciousness, cosmology, alternative medicine, dreams, climate, permaculture, alchemy, shamanic studies, oracles, astrology, crystals, hyperobjects, locutions, and subtle bodies.

Cataloging-in-Publication Data for this title is available from the Library of Congress

ISBN 978-1-64411-845-0 (print)
ISBN 978-1-64411-846-7 (ebook)

Printed and bound in the United States by Lake Book Manufacturing, LLC
The text stock is SFI certified. The Sustainable Forestry Initiative® program promotes sustainable forest management.

10 9 8 7 6 5 4 3 2 1

Text design and layout by Priscilla Harris Baker
This book was typeset in Garamond, with Futura, Gill Sans, and Merel used as display typefaces
The following images are courtesy of Wikimedia: plate 1 (CC0), plate 2 (CC 3.0), and plate 4 (CC BY-SA 3.0).

To send correspondence to the author of this book, mail a first-class letter to the author c/o Inner Traditions • Bear & Company, One Park Street, Rochester, VT 05767, and we will forward the communication.

Scan the QR code and save 25% at InnerTraditions.com.
Browse over 2,000 titles on spirituality, the occult, ancient mysteries, new science, holistic health, and natural medicine.

In beloved memory of each of our ancestors that
weave generations of connection, remembrance,
strength, and wisdom through words, stories,
and universal sound. May our souls
listen deeply and sacredly.

To Michail Remennikov, Reyza Remennikova,
and Anna Gorodetskaya.
To Richard Grossinger and the family of
Inner Traditions International.
To Dave Robinson.
And to you, dear reader.

oᴠᴠᴠo

Behind each letter, there is a vibration. The collective vibration introduces a word that shares meaning. Within each meaning, there is a story. That story underscores life, and one that was created with hope and love. The space in between lies the hidden, infinite potential of the moment.

oᴠᴠᴠo

Contents

Foreword by Richard Grossinger   ix

Introduction: Activating   1

ᏮᎢᎢᎵᏮ

PART 1
DNA, Healing, and Evolution

1   What Is 12-Stranded DNA?   12

2   The Art of Memory and Generational Healing   20

3   Devolving to Evolve   34

4   The Genetic Archetype and the
Dodecahedron Form   48

5   Revealing Our 12-Stranded DNA by
Pythagorean Gematria   62

6   Molecular Gematria   69
DNA Strands and Morphic Evolution

7   Water Memory and Mysteries   83
*Speaker, Shape-Shifter, and Heart
of the Dodecahedron*

8   Water Frequencies   92
Responding to Music, Mantras, and Intention

9   Sound Alchemy   99
The Healing Power of Vibration

PART 2

Breath, Animals, and Angels
Heal, Activate, and Evolve Your Dodecahedral DNA

10   Draw Breath and DNA Consciousness   112

11   Dreamtime and the Ancestral Realm   128

12   Composting Grief   143

13   Animal Medicine and Holism in the New Age 154

14   Divine Codes, Human Origins, and
         Angelic Frequencies   180

15   Angelic Ferocity   191
         Dodecahedral DNA and the Language of Angels

PART 3

Advanced Practice
Back to the Drawing Table

16   Gematria Meditations   209

17   The New Sound Frequencies   221
         Sound Exploration of DNA in the New Age

Epilogue   237

Notes   242

Bibliography   249

Index   261

About the Author   271

Foreword

Richard Grossinger

I began working on Ruslana's book—from here on, she will be Ru—
on May 21, 2022. I am pondering how to present her work in such
a way that you, her readers, get the same supersonic—she would say
"aerodynamic"—boost and personal transformation out of it that I
have. This is Ru's first book and merely hints at the incredible teachings
that will come from her in her future texts and transmissions.

I was skating during the public session at the indoor ice rink in
Portland, Maine, today, thinking about this project, when the last digit
of reflected time in a glass partition looked like a "5" but was actually
a "2." That caused a momentary confusion about how close the session
was to ending, and in that moment a universe arose, flickered, dissolved,
and delivered this foreword.

A "5" that is actually a "2" touches on the alphanumerical mystery
and code at the heart of *Activating Our 12-Stranded DNA*. The rink
Muzak was particularly tinkly and, with the murmur-chattering of the
skaters and giant fan, the "5 to 2" fluctuation synesthetically recalled the
Bon Iver album *22, A Million* in which the bands are not quite songs as
much as sounds, and each is titled by an alphanumerical trope with other
characters thrown in suggesting algorithmically generated passwords.

I have always played with words and numbers, looking for correspondences in the world, not necessarily magical ones. It goes back to
childhood when I routinely reversed words, looking for palindromes,

playing Scramble or Jumble to see what other words were hidden in them, and sometimes noted high Scrabble scores, for instance, on billboards and license plates. I was not alone. When schoolmates and I took the crosstown bus at 96th Street in New York City in the late 1950s, we asked for transfers that allowed a free subway ride to the north city. We played with the serial numbers on these paper slips, racing each other to get ours to end up at 10 by any combination of mathematical operations including square roots.

Ru takes these sorts of operations to a much higher mystical, Pythagorean level at which point they shed light on your own world path and the future of the world itself. The Pythagoreans have performed this way since ancient Greece. The world is filled with untitled oracle decks. Earlier this year, I went on a hike in Acadia Park with Ash McKernan, a practitioner of divination and self-declared initiator of the "Wyrd Way." During our hike he found a piece of torn cardboard with the drawing of most of a snake, an unidentifiable piece of metal, a tiny jigsaw from a puzzle, and a few other objects whose identities I forget from ground litter. He tied these together with a raven that followed us for a while from tree to tree. He read his own oracle for the coming weekend about which he was trying to choose from among varying options.

Ru does something similar to you. She changes the way you look at the world, the way you hear sounds, the way you look at words, images, sigils, and signs. It is not superstitious or restricting like some oracles; it is expansive, filled with hope, humor, and love. She is also the most affable healer you'll ever meet. I think of her as a waveform of a future evolved Ru merging with her social persona, a visitor from outside the modern global crisis. At once, she has absolute boundaries and is transparent and deep as the sky. Her transmission sometimes seems Blakean, recalling his "A Little Girl Lost:"

> *Children of the future Age,*
> *Reading this indignant page;*
> *Know that in a former time,*
> *Love! sweet Love! was thought a crime.*

Or as a Blakean microbiologist "The Tyger:"

> *And what shoulder, & what art,*
> *Could twist the sinews of thy heart?*
> *And when thy heart began to beat,*
> *What dread hand? & what dread feet?*
>
> *What the hammer? what the chain,*
> *In what furnace was thy brain?*

Blake follows with the ultimate question, particularly in this age of CRISPR and 12-stranded DNA:

> *Did he who made the Lamb make thee?*

Ru is uniquely qualified to address that, for she scries not only helical building blocks of lion and lamb but their oversouls, sigils, and messages. *Activating Our 12-Stranded DNA* is as Blakean as it is biological. She also elucidates Michel Foucault's inaugural message in *The Order of Things*:

> [N]ature, itself, is a seamless fabric of words and marks, of stories and letters, of discourses and forms. . . . There can be no commentary unless beneath the text that one reads and explicates, runs the sovereignty of an original Text.

A few months ago, my wife and I crossed Portland's downtown green, through a homeless encampment toward Saturday's farmers' market. A flock of huge Canada geese were landing, pecking at the encampment's scraps, pulling up dead grass, and stopping a major traffic flow as they crossed the street toward the market, not in air but by foot. A cell call led Ru to tell me about geese and how they are generally a sign that you need "to put your foot down."

Instantly two prior "geese" frames lit up, and now every time I see geese, or every time their oracle is relevant, I think about whether I am standing up for myself enough.

After you have passed through Ru's wand, all of the animals in the world take on a similar divinatory quality and hold a meaning for you, but (again) only when it is relevant. No obsessive-compulsive disorder here, just grace.

When our cat Tawny disappeared for a few days that July, Ru sent an email:

> Start every ritual with gratitude and calling on your angels, allies, guardians, and protectors for support. On a sheet of paper, write a note addressing it to Tawny, requesting her to come back. Write it as fondly as you like (example: Sweet Tawny, please come home, safe and sound. Your presence is missed dearly. We will wait with open arms.) If you have incense, light it and circle the sheet 7 times clockwise and counterclockwise each, or if you have a candle that will work too. This is a meditative and vulnerable practice, be as present as you can. Leave the ritual open until the candle has burnt on its own, or the incense has dissolved and you can't smell it anymore and put the note into the nightstand drawer next to you.

As the first editor of this book, I told Ru that her voice is like a pure yellow yarn running through the text, but sometimes she loses her own thread and drifts into psychobabble or kitsch. She forgets to track the pure yellow of her thoughts into the language of the book, and it browns or grays out into the fuzz of common parlance, losing her individuality. Yet it's there behind the words—listen quietly.

Sometimes it's a consequence of her ESL (English as a second language), which comes out only in her writing. Her spoken verbs and overt and covert syntax are impeccable, but when she has to key them, brain to fingers, a different part of her brain that was not originally programmed in English takes over, and an older, deep syntax rules her logic strings. This can be fixed superficially but not corrected. The yellow yarn is there, but you have to listen for it. On reading this, she wrote me:

> One thing about the "ESLness"—by the natural timeline, perhaps even divine, I was four years behind learning how to read and write because of the timeline when my parents immigrated, getting set up

with a home, and getting me into school. I didn't attend kinder-garten, spending that year with my grandmother at home instead. I entered first grade without knowing or understanding English. I knew a variety of Russian, English, and Hebrew words but making sense of it all was a challenge. Instead, I listened to the sound and observed people's actions and emotions. My parents couldn't teach me English because they were learning it at the same time. It turns out that kids can actually fail first grade—and I was a statistic. My dad arranged somehow (I don't know the full story) for me to keep going into the second grade. I remember my teacher, Mrs. Amy Vanderpoel. I was scared coming into class on the first day because as a child, I already understood that it was like a second chance and I better not mess up. She held my hand and promised me she was going to be there by my side each step of the way, and she was. Because of her lessons with me on the fly, I was able to catch up. From then on, I graduated high school in the top ten of my class.

At the end of each year in elementary and middle school, there's a school-wide spelling bee. I was able to become a finalist (not necessarily the champion) each year beginning in the second grade because of sound and imagination. I was stunted by the word "chauffeur" in second grade. In my mind, I thought there was only one kind—shofar, of course! In the sixth grade, I lost to Chung Yi because I was terribly embarrassed to spell out the word a-s-s-a-s-s-i-n-a-t-e to honor and respect my family (two "ass" fragments). I intentionally spelled it a-s-a-s-i-n-a-t-e. My face was so red and I was disappointed as I was spelling it and pulling out my white flag.

Her second editor, Chanc E VanWinkle Orzell, took care of much of that, pulling a blue and orange text out of gold fleece.

Ru is also a consummate sound healer, a musician, an operatic and pop singer, a potter, a baker, a florist, and a former pharmaceutical drug developer on the science mainline. When she writes about DNA, she is writing from direct experience as a handler of its elixir and its enigmatic twin gyres and gematrias. When she finds twelves in water or reads fish talk in aquarium brew, she performs as an astute chemist and crystallographer. These messages are there at some level. When she trains and conducts sound, she is doing so as a one-woman experimental band. The glossolalia lingers long after its vibrations fade. The sounds are directed toward cellular, 12-stranded helices where they babble metaphorically in CRISPR and junk DNA.

A very raw precursor of this book was submitted to Sacred Planet Books in April 2022. It read like a combination model's portfolio and old-fashioned *People* magazine and influencer pitch. It featured Ms. Remennikova as a triathlon athlete, participant in Ironman competitions, and former food truck operator, restaurateur, tennis coach, and tennis player of rising rank. I had to look behind all these manikins to see the sage medicine woman, but she was there—she can't be hidden—enough for me to say, "Let's do your real book." The editing began.

Ruslana was born in Kyiv, Ukraine, and came to the United States when she was two. Her father didn't like the vibe of Vladimir Putin at the time. As the former KGB agent began to foreshadow his intentions about Ukraine, Michael Remennikov moved his family to Virginia where he imposed a paragon of excellence that was sometimes excessive but always based on the reality that you can be forced to flee at any time, any place, so you better learn your stuff while you can and earn your place in society. It is why I have teased her, I hope not blasphemously (for she is *not* in a war zone herself, though many of her friends and relatives are), that she is at times like the whole Ukrainian army. That's how quick a study she is and how fast she plows through terrain. Just before we began editing she wanted to see the American West. She and her partner left Virginia and took in the Dakotas, Wyoming, Idaho, Montana, the Southwest, and Colorado in two and a half weeks, including getting back home, and they did not travel trivially, as the animals and portraits in this book show. They were a regular Lewis and Clark Expedition.

I also anointed her with "angelic ferocity," which made its way into the book. In an earlier version of it, she told a story about a college exam:

Dr. Shillady made our exams simple: there were five questions on each one, but each question required multiple pages of theorem proofs—translations of chemical mechanisms and pathways into math and physics. It looked like something you would see in *The Matrix*, floating numbers, symbols, squiggly lines, carets, and arrows. Expressing my respect to Dr. Shillady in a proper Ukrainian manner, I showed up to take the exams in a white suit, black heels, red fingernails, and groomed wavy hair. It was my ritual of honoring him, the course, the ancestral knowledge, the trial, and my own hard work. The first time Dr. Shillady experienced my unexpected ensemble, he was surprised. As I turned in my exam, he made an announcement in the middle of the exam, "Ladies and gentleman, Ms. Remennikova is leaving the building."

I keep coming back to the notion that sound healing is Ru's encompassing trope. Everything is a mark in silence, and every mark serves divine creation. Ru was quite miscast in the pharmaceutical world, but it was an essential phase in her individuation and metamorphosis. This book is another phase. It is less a document than a process, a journey you take with her and yourself. There is no formal finale. The outcome is the use that you make of its text. That is true of all texts, but this matrix is really wide open, mixing gematria, animal guides, and 12-stranded DNA. If you open your beliefs to it, a Ru wind will take you to your own truth mystery.

RICHARD GROSSINGER is the curator of the collection Sacred Planet Books, published by Inner Traditions, and is the author *Bottoming Out the Universe* and *Dreamtimes and Thoughtforms*.

INTRODUCTION

Activating

DNA! The abbreviation has an electrifying effect. It conjures a procession of images across the mind's eye, bearing impressions that range from a twisted ladder to a golden spiral. I am struck by the beauty of the mandala revealed by a full cross-section view of DNA, though I cannot capture it in words. Its structure animates Pythagorean orbits, vibrant shapes longing to be "heard," and the power to dovetail emotional connection and engagement with the exciting and inspiring building blocks of life that make up our mind and body. The initialism of DNA itself shares a gematria (which will be covered throughout the book) with the classical Greek* term for "magic," μάγια (mágia), which means "enchantment," "charm," and "wizardry." You carry this wealthy estate, fortune, and wisdom.

While the global pharmaceutical market pinnacled at 1.6 trillion US dollars in 2023, the absence of the deliberate participation with one's own biochemical unique nature leads to the demise of mindful evolution. I would cite lack of know-how as a cause. Evolution is a divine art. The potential moving from one aspect to another is felt as

---

*In the English language, both "Greek" and "Hellenic" are used interchangeably to indicate the people and culture correlated to modern day Greece. However, Greeks refer to themselves as *Hellenes* or Ἕλληνες. Reference Chapter 5 on the history of gematria and how the language, culture, and values of Hellenism shaped the innovative works of the modern world.

it is in music, where one melody can be transposed from one key to another.*

The book currently resting within your grasp is not a science text, nor is it written as such. It is meant to act as a vibrant guidebook, charting out a dynamic pathway toward the elevation of your consciousness and spiritual being. DNA is beyond a static blueprint. It specifically captures the macrocosm that the microcosm models. This macromolecule, the key to our destiny, has intrinsic value apart from any conventional or esoteric interpretation we make of it.

It is accurate—and more than merely accurate—to say we *do* have a deep influence on our condition when we focus our will and desire on our cellular being. Perhaps, instead of asking what our genes can do for us, ask how we may honor DNA.

THE STORY OF THE DOUBLE HELIX DISCOVERY

Nearly seventy two years ago, on April 25, 1953, *Nature* published three consecutive articles communicating the works of James Watson and Francis Crick, followed by Maurice Wilkins, succeeded by Rosalind Franklin and Raymond Gosling. James Watson and Francis Crick assumed pioneering titles for the double helical model's debut. However, there is a rather confronting, pertinent question to ask in this moment

---

*A note regarding *completing our human evolution* in this book's subtitle: evolution is a continual process. Ultimately, there is no final form of an organism based on the factors that influence its evolutionary cycles including internal changes, climate, migration, vegetation, and so on. What the title means to propose is that the dodecahedron is a universal gateway that can be utilized to approach an archetypal example of genetic composition as a whole *as we continue* our evolutionary inquiry of the human organism. The word *complete* is reconstructed from Proto-Indo-European *pleh*, which means to fill. In ancient Greek (Proto-Hellenic) etymology πλήθω (plêthō) is a transitive verb that comes from the root *pleh*, meaning to go through a period of time, to be and to become. The intended use of the word *complete* in the book's title means to examine the dodecahedron as an evolutionary model, which acknowledges the numerous iterations that comprise its physical and ethereal form. Each iteration honors its next version, evidence that DNA and evolution exhibits a recursive system or fractal representation of creation. The meaning "to be" or "the becoming" is the intended application of *completing our human evolution* for the title and book.

in history and psychological perspective. Who is the actual face behind the groundbreaking conclusion that the DNA molecule exists in the form of a three-dimensional double helix? And, why is the recollecting and remembering of this historical segment in solving a biological riddle critical to reassembling our orientation with the most crucial landmark for humanity, our DNA?

Around the same time Rosalind Franklin arranged for a three-year research fellowship in John Randall's lab accompanying Maurice Wilkins at King's College London, James Watson attended a talk in Italy and learned about Wilkins and X-ray diffraction data for DNA. His interest in DNA was piqued: "Suddenly I was excited about chemistry. . . . I began to wonder whether it would be possible for me to join Wilkins in working on DNA."[1] Watson recruited Francis Crick, whom he met at the University of Cambridge working in the Cavendish Laboratory.

In the beginning of 1951, Rosalind Franklin's discovery of the properties of DNA (using a DNA sample of a calf's thymus) were not only profound, they were also higher in data quality than Wilkins's. At the time, Franklin was assigned as an assistant to Wilkins, who was pioneering work in DNA. As a consequence of a lack of communication about their respective roles from John Randall, the director of the biophysics unit, the DNA problem was reassigned to Franklin as her sole responsibility while Wilkins went away on vacation. This caused a clash of personalities within their working relationship.

In her X-ray diffraction imagery, Franklin discovered DNA could exist in two forms, wet and crystalline. To delegate research assignments and responsibilities, Randall apportioned their work on DNA. Franklin chose the wet DNA and Wilkins stuck to the crystalline form. The crystalline DNA was less rich in information than the wet, long and thin DNA fiber.

At a lecture that was attended by Watson at King's, Franklin proposed that the results of her research suggested a helical structure (which must be very closely packed) containing 2, 3, or 4 co-axial nucleic acid chains per helical unit and having the phosphate groups near the outside.[2] This was sixteen months before Watson and Crick published their description of DNA based on Franklin's X-ray diffraction photos.

While it became clear that the crystalline DNA is a helix, Franklin was unconvinced that the wet form was helical. With her assistant, Raymond Gosling, she produced the world's best X-ray images of any substance to date. Franklin then went on to reconcile her conflicting data that both DNA forms are helical. In 1953, she wrote three manuscripts, two were on the crystalline form of DNA (A-DNA), and one was on the wet form (B-DNA). Her manuscript on A-DNA was received by *Acta Crystallographica* on March 6, 1953, a day before Crick and Watson completed their model on the wet form.

As a scientist who worked on the same subject in another laboratory, Watson visited King's to inform Wilkins and Franklin before Linus Pauling and Robert Corey discovered the mistakes in the preprint of his proposal for the structure of DNA—a scheme of a triple helix—and outrace them. During his visit, Watson had his infamous altercation with Franklin. He upset Franklin by saying she didn't understand how to interpret her own data. Watson retreated to Wilkins, who showed Franklin's X-ray photos to Watson out of sympathy.[3] Following an unauthorized, nonconsensual use of her unpublished crystallographic calculations, Watson and Crick developed the double helix model of DNA. *Nature* published their model in April 1953 with only a footnote acknowledging "We have also been stimulated by a knowledge of the general nature of the unpublished experimental results and ideas of Dr. M. H. F, Wilkins, Dr. R. E. Franklin and their co-workers at King's College, London."[4] Neither James Watson nor Francis Crick mentioned her contributions during their Nobel Prize acceptance speeches, though Wilkins did. In Watson's book *The Double Helix*, he admitted "Rosy, of course, did not directly give us her data. For that matter, no one at King's realized they were in our hands."

After completing their model, Watson and Crick offered Wilkins to be a co-author of their paper. He declined it, as he realized that he had no part in the construction of their model. The original contributor, Franklin, wasn't nominated for a Nobel Prize, yet her work led to Francis Crick, James Watson, and Maurice Wilkins being awarded one in 1962. Franklin's own rewards were few. A young life of brilliant and continual productivity came to an end in 1958 at the age of thirty-seven. The structure of the DNA model was not even fully accepted during her lifetime. Posthumously, this book carries the flame that inspired her work.

I wonder how the advancement of the double helix structure would have proceeded if she had lived longer. Of course, Franklin wasn't the only contributor to the discovery. Johann Friedrich Miescher, Albrecht Kossel, Gregor Mendel, Alfred Hershey, Martha Chase, Erwin Chargaff, Michael Creeth, Linus Pauling, and Phoebus Levene are among a number of giants who dedicated their lives to excavating DNA. However, would Rosalind's relentless curiosity entice her to explore the unseen aspects of the genetic molecule? The double helix is just the beginning of our path to biological enlightenment.

In this book, I am going to take you on a journey. It is the sort of journey in which the very small becomes large and reveals universes within it. Within these universes, sounds, colors, and shapes turn into one another in synesthesia. But we cannot explore these phenomena overtly. I am going to approach space-time through sound, DNA strands, liquid water crystals, animal and angel guides, lucid dreams, and ancestral grief. Each is a proto-cell, a model system, and a world. I am going to string these droplets together by their vibrations and convert their elements into pathways to each other. Through these tools, we will explore the reality of our 12-stranded DNA.

Deoxyribonucleic acid, DNA, is the biochemistry of our genes. The word *gene* invokes different worlds of contemporary culture: politics, biotechnology, medicine, lifestyle. The DNA meme has become a timeless icon, symbol, and force by itself. It tells us many different stories, ones that are true, ones that are probable, ones that depend on specific conditions, and ones that are unreliable. As a power and a prop, DNA leaves us with ambiguity, tautology, and obscurity.

"Gene talk" is institutionalized and marketed through inanimate objects across international networks of social, scientific, and commercial exchanges. Just look at the commoditized overlay: DNA perfume by Bijan is inspired by "the power of heredity;" Brooklyn's DNA Footwear shoe brand is "street-chic meets urban sophistication for your Mind,

Body & Sole;" a vintage Toyota ad illustrates a "great set of genes."
News media and entertainment platforms tag the word *gene* or *DNA* at
the end of a concept for drama, humor, or commodity. There are cool
kid genes, obesity genes, rock-star genes, depression genes, bad genes,
lazy genes, sexy genes, six-pack genes—the list goes on.

The company 23 and Me won the brand war for hereditary analyses
not by a superior product but by superior marketing. In a three-word
meme, founder Anne Wojcicki and her partners caught the personal
nature of the gene, the numerology and twin spiral of 23, and the power
of numerology itself. That is, the critical necessity that numbers provide
is a means to relate, communicate, and understand the external world
and interpret cosmic creation.

The primal essence of genetic archetypes carries a sort of innate
contagious magic. Even scientists speak of genomes archetypically,
as the "Delphic Oracle," the "Bible," the "Holy Grail," and the "God
code," the last one in part for its presumed potential to confer genetic
immortality. It is divine enough that it confers life.

To assume that DNA's function is only to synthesize proteins and
regulate gene translation is reductionist and myopic, missing its multifaceted caliber. DNA can be influenced, queued, guided, interpreted,
and raised vibrationally by our thoughts, dreams, aspirations, meditations, and spiritual awareness. Its mystical construct tells a story about
the most sacred meaning of life that parallels other philosophical, theological, cosmological, and psychological narratives.

Our journey, then, will be to explore a new science, leaving behind
the traditional double helix deified in the scientific canon and popular
culture and seeing DNA through a new lens. This book offers techniques, wisdom teachings, and meditations for healing your DNA and
activating your genetic evolution. We will learn a new language using
sound, water, animals, dreams, the soil, ancestral and angelic communication with calculated applications of divine geometry, physics, chemistry, and gematria.

We will come to understand the underlying order of the 12 strands
of the dodecahedral DNA, focusing on the natural archetype associated
with the number 12.

Inspired by Plato's hidden codes and methodology, I reveal decrypted phrases and messages regarding our "12 strands" and the underlying dodecahedron using a Pythagorean approach. With this foundation, we will explore a variety of techniques, codes, and formulas, bringing forth the elusive and unseen. My techniques highlight the ways in which we can use our natural abilities to heal our thoughts. Think of it as a rite of initiation, your birthright and sacred dowry. The ocean of unconsciousness to which everything is linked—events, actions, thoughts, feelings, willpower, imagination, prophecy, and clairvoyance—is accessible to all through intentional methods of practice including meditation and sound healing. These are where the mortal and immortal meet and establish your autonomous relationship to your DNA—one that will bring you clarity, empowerment, and compassion.

As you breathe consciously, your breath will spread new understanding and dimensions. You will recognize and celebrate your individuality and irreplaceability within the collective. We need each other as much as the universe needs us, hence our life's appointment now and here.

In the coming period, the dodecahedron is the heart of DNA, the new icon of human design with the Universe, the so-called God code, and a spiritual access point of the New Age.

This is our journey.

Are you ready?

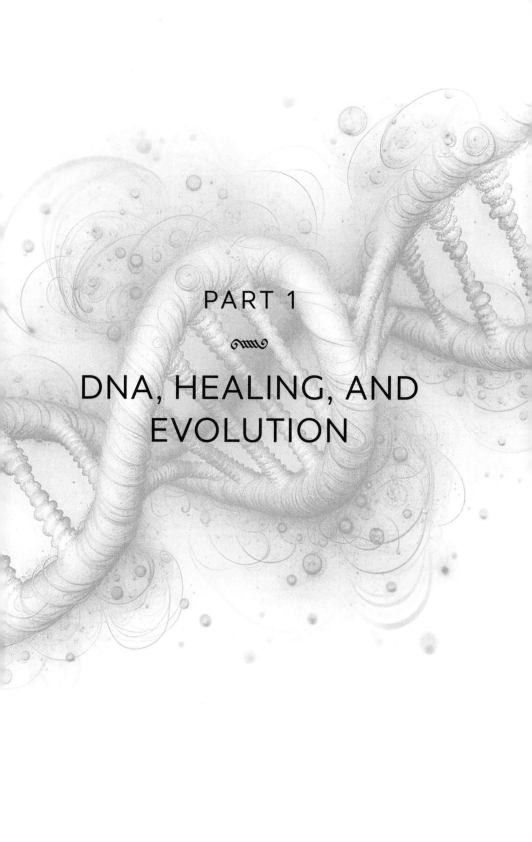

PART 1

DNA, HEALING, AND
EVOLUTION

Introduction to Part 1

People often get stuck at the vibration of *feeling* stuck or sick. Similar to fear, feeling stuck is a form of illusion. There is only the present moment, and the next. The space in between is our sovereign relationship to incarnation—time keeps us embodied. Even with our genetic complexity, we can alchemize our thoughts and instill our feelings with various energies. Our archetypal DNA provides a repository capable of overcoming emotional stumped spots in multiple ways. A key idea in this book is how we can change our DNA. Cephalopods like octopuses do it unconsciously in the same way that they change color, shape, and acclimate to their environment's temperatures. We can try to make this process conscious at the level of attuning our vibration.

Our DNA is at once our link to genome memory and an outward voyage through an inward library of potential. This is how the aetheric plane operates on the physical plane. According to Rudolf Steiner and other theosophists, that's how we got onto this dense physical planet—from the astral plane. We are capable of making shifts that create landscapes and opportunities, if not whole planets all at once— that's a collective, generational feat—but creation at our own scale. We make ourselves more intentional by the words (sounds) we say and, correspondingly, the sounds that we listen to. We treat our dreams and hypnagogic and hypnopompic states as oracles that conduct messages, beliefs, and wisdom codes. We open an ongoing relationship with nature's innate magical properties. We become listeners to all these different unseen languages. This book is an aid for recognizing symbols and meanings as subtle languages to aid your evolutionary journey.

My method to activate your 12-stranded DNA is more than a meditative practice. It is more than a consistent eating, sleeping, and movement regimen. It is beyond interpreting external frequencies. It is about falling in love, over and over again. It is about the celebration of your ethereal voice. It is the way your purpose and soul constellate. It is about embracing the relationships from your past and future with a desire to appreciate that it all *meant something*. While biotechnology explores this concept through robots, pharmaceuticals, or other artificial means, we are able to propel divine genus through an organic and sovereign synapse.

A point I want to make clear is that deep aetheric and higher changes, changes on a level of prana and vibration, do not show up directly either in laboratory scans like X-rays, MRIs, or CT scans. They are on an energetic level and manifest through other biochemical processes. You can no more track the effect of a mantra than of a homeopathic medicine. Too many people think that if you do a particular exercise or practice yoga or qi gong devotedly, a biological change will show up in a lab test. It doesn't work that way. You do yoga or participate in a sound bath or conscious breathing to change your overall vibration. When your vibration changes, you change. It gets recorded in DNA in Lamarckian fashion, something that Darwinian science says can't happen. But my book is about speaking to primal code in its own language—vibration—and changing. What I mean to propose in part 1 of this book is a mix of alchemy and new science.

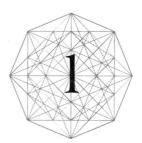

What Is 12-Stranded DNA?

Deoxyribonucleic acid (DNA) is a technology created by nature through millions of years on Earth. It is the natural transfer of a higher system of organization released through an aetheric portal of spiraling clouds of stars within spiraling galaxies. DNA is a complex, reprogrammable language that can be influenced by light and sound, natural and primordial factors, to transcribe its values and transmit them into life and creature evolution. But DNA is not just an informational construct; it is more intelligent than its distilled scientific intellectual alias.

DNA is consciousness evolving with matter. Why is that true? It's true because consciousness existed before matter in this universe. Consciousness is information becoming matter. You cannot get a crystal as complex and intelligent as quartz or diamond, let alone DNA, without it being informed by the universe itself. When you have a belief and thought, that's because the primordial DNA, operating through your own DNA, is such an intelligent molecule and a direct participant in creation. We see clues of this in the animal kingdom, where creatures respond to unconscious biological information, such as environmental changes, through genetic interactions, altering the generational passing down of their genes.

When I first saw mold growing on trees as a little girl, I realized we live in a mystical world that is beyond the information passed on in traditional education. I appreciated the models of science, but my next usual

question after a newly introduced concept was, but why? Fortunately, I was a quiet, introverted child; otherwise, I would have driven my educators mad, which happened anyway as I began my career as a research pharmaceutical scientist and developed the confidence to use my voice—squeak! howl! purr! *But the same question led me to the internal worlds that we each carry. The complex systems of information that we outwardly express exist latently in a dormant morphogenetic state mirrored in the external world. That's what the mold was telling me. We are much vaster than we can grasp, just as a slime mold is. It may not look lovely, and it is not petalled out for a picnic of bees like a rose bush. However, when we understand its origins, behaviors, and mechanisms, we see that it is not only quite gorgeous, but not so different from us in the end. We are DNA, epigenetics, information, agency, and growth at once.*

THE STRUCTURE OF DNA

DNA is composed of building blocks called nucleotides, consisting of a base, a deoxyribose sugar molecule, and a phosphate group. The bases— adenine (A), thymine (T), guanine (G), and cytosine (C)—each have one of two possible geometric ring formations, divided into purines and pyrimidines. These base pairs have been found on carbonaceous meteorites—in places where life as we know it cannot form; they are primal, star-cloud-born shapes. A base pair refers to two bases that combine to form a "rung of the DNA ladder" between two strands. As life- forms come into being, they recognize, copy, and transmit their 3-D templates in paired helical strings that look like a long spiral staircase. (See plates 1, 2, and 3 in the book's color insert.)

THE STORY OF THE DOUBLE HELIX DISCOVERY

How was the double helix discovered? If you think it is only the model announced triumphantly by James Watson and Francis Crick in February 1953, you are only right about one level of DNA and one aspect of its discovery in a particular form. You are right about the 1962

Nobel Prize in Physiology and Medicine but not about the multiform and fluid nature of DNA, which is more like a hive of shape-shifting bees than a pinup of heredity for a blue-ribbon committee.

While most of the world believes that Watson, Crick, and Wilkins discovered the double helix, I laid out a more accurate history in the introduction, which describes the significant contribution of Rosalind Franklin. If Franklin didn't exist, Watson and Crick's understanding of the DNA helix, hence our own, would be very different. If prizes are called for, it should have been a four-way, not a three-way, blue ribbon.

Even presuming Watson, Crick, and Wilkins were the discoverers of that format—a double helix carrying heredity's information—that helix is only a micro-material form of a much vaster construct with a twin-helical structure, which itself is a gateway to an even vaster flow of information. In the Hindu version of aetheric energy channels, the prime *nadis*, called *ida,* and *pingala* spiral about each other in the same shape as the DNA helix. That alone should give you pause about what biotech laboratories are working with: divine, sacred dancers and their tiny tridents. As in many technological adventures, scientists only touch the concrete version of an archetypal form, the one to which the methods they use lead them.

There are two reasons why I share the story of the double helix discovery with you. The first is the essential role of water: the water molecule underwrites the discovery of the double helix. It is not solely a matter of hierarchical development: water being simple, DNA being complex. Their organic hydrogen-oxygen bonds are spatially intercalated. Water provides the cornerstone for the dodecahedral archetype behind DNA that we will explore in this book. Second, the Watson-Crick adventure story both shows and conceals the aetheric roots of the double helix. While their structure trailblazed science and medicine and affected the way that we understand evolution as a forward-facing information flow, it's not the whole DNA picture by a long shot. Their misconstrued dynamics of the double helix disguises its *actual and deeper* dodecahedral structure.

In this book, I, a modest acolyte in the Watson-Crick temple, both honor and move on from their model. The Watson and Crick double helix is, to a certain extent, a key to our 12-stranded DNA. However, my focus is on the dodecahedral imprint as the foundational basis of cosmic and genetic exploration and angelic communication—I call it trans-DNA messaging. Of course, Crick and Watson were not trying to do what I am—they were pure scientists, not musicians or healers—and their aspirations were limited to the biophysical easel. I take their work with due respect, but I have greater respect and uses for the repetitive universal mechanism of hidden codes, geometries, and gematrias inherent in DNA. DNA understanding isn't a linear process, nor is it limited to the purviews of a laboratory, medical facility, or clinic.

DNA IS NOT A FIXED STRUCTURE OR SHAPE

Insofar as our own biology attunes to the vibration of the universe, DNA is not a fixed structure or shape. In 2020, scientists at the Imperial College London discovered a stable 4-stranded DNA structure created by normal cellular processes. Zoë Waller, associate professor in drug discovery, Pharma and Biochemistry at the University of East Anglia, says the quadruplex structures form part of normal DNA regulation and function, and that our view of the DNA double helix may be out of date: "We presume that this is the normal, native state of DNA, but this work is another exceptional example of mounting evidence that DNA is not a fixed structure or shape."[1]

My work goes further. I propose an archetypal 12-stranded DNA, as revealed through the mysteries of palindromic DNA sequences and gematria.

DNA LANGUAGE
Palindromes, Gematria, and the Five Platonic Solids

Palindromes

The basic spiral structure of DNA is composed of two linked strands that are antiparallel, hence they are parallel to each other but oriented in opposite directions. One end of a DNA molecule is referred to as five prime, 5', and the other end is referred to as three prime, 3'. When complementary double strands read the same in both directions, either from the 5' end or 3' end, the nucleotides create a palindromic sequence. For example, the sequence GACATGTC (or 12324143 if assigned numbers on one strand of DNA is considered a palindrome because the sequence on its complementary strand is CTGTACAG (34142321).

Palindromes are everywhere: listen to crows chanting them or watch palindromes form and dissolve in clouds. The Hellenic phrase for *the palindromicity* is η παλινδρομικοτης, which contains a palindromic gematria number 1001. Where DNA's palindromes meet alphanumerical palindromes is a depth of nature beyond our capacity to plumb, though it is created in and by our consciousness. It is the alphabetic, numerical, palindromic basis of all organization and information. Gematria is its own occult science, and I will explain it gradually as I go along, so if you are unfamiliar with it, be patient.

Nucleic palindromes exist throughout the human genome and are a widespread mechanism in life that, for instance, equips bacteria with a heritable adaptive immune system and genetic memory. Palindromes are clearly more than children's *toots** or adult word games; their abundance, frequency, and location speak to their innate functional roles. They affect how genes express, regulate, and replicate information. Palindromes stimulate nearly 83 percent of deletions and insertions during DNA replication and other genetic operations.[2] If this is word play, it is play of the most profound and creationary nature.

The palindromic system I use to interpret the 12-stranded DNA reflects the same spatial arrangement and orientation seen in genetic palindrome sequences—symmetry, numerical significance, and complementarity—while using the earliest forms of scientific modeling: gematria and geometry. My system treats palindromes in the way traditional science does, as a bidirectional language and deep syntax of DNA.

Gematria

Gematria is a key tool that I use in this book to decrypt phrases to a numerical value, which I then compare to the numerical value of a separate

---

*Toot is coincidentally a palindrome. By context, the word *toot*s is applied to indicate that palindromes are not a child's play. They are a property of nature and animated human experience.

phrase. If the numerical values of each phrase are equivalent, the phrases have meaning and relevance to each other. Understand right away that I share with you the view that gematria is ridiculous: a combination of what appears to be meaningless coincidences, rigged data, and magical thinking. Guess what: it also works. Gematria is one of the tools of angels, guides, spirits, and archetypes to reach us here in the material density in ways that few other things can. Numbers are tried with technology, but they are generated by gods. That is, they originate in cabalistic and divinatory systems of all kinds. It is possible that the Greek oracle at Delphi used gematria. In one form or another, its pre-algebraic seeds have been used by indigenous peoples across the earth, sometimes in the form of plant shapes, star arrangements, eggs (the origin of zero), and the calls of birds.

The Greek author and mathematician Eleftherios Argyropoulos used Pythagorean gematria in order to provide—"prove" in the terms of a time more receptive to angelic interference—the metaphysical basis for the correspondence between the five regular solids and five cosmic elements. The nature of this proof demonstrates that in each word, phrase, or sentence there correlates a unique and equivalent number to two separate words or phrases. The identical numerical sum expresses the deeper relationship and significance between two such ideas or concepts. More significantly, the relational distinction of this mathematical value shared between such essences is an abstract, or element, of an eternal and imperishable construct of reality. The intelligence that underlies Pythagorean gematria will be discussed further in chapter 5.

The Five Platonic Solids

In the sixth century BCE, the Pythagoreans discovered five solids, which have the following properties: each of them is constructed by using the same regular polygon, and all the three-dimensional angles in each are equal. Later these solids were called Platonic solids because Plato and his students explored their geometric properties in detail. Plato was the Einstein of a golden era in ancient Greece, and his work laid the basis for Western culture and its sciences. We will look more closely at these shapes and their relationship to the dodecahedral, 12-stranded DNA in chapter 4.

"THE GOLDEN RACE"

Pythagorean Divinity within the Dodecahedral DNA

12-stranded DNA is a result of my interpretation of God as found in the form of Pythagorean gematria. I treat the meaning of "God" as a secular numerical entity, yet evoking a compelling trifecta: cosmology, theology, and symmetry, the latter beginning with amicable numbers. Amicable numbers are two distinct natural numbers that relate in such a way that the sum of their individual divisors is equal to the other, and vice versa. Amicable numbers were introduced by the Pythagoreans, who believed them to reveal important virtues relating to friendship and justice. The first five pairs of amicable numbers are: (220, 284), (1184, 1210), (2620, 2924), (5020, 5564), and (6232, 6368).

One way to consider them is that amicable numbers represent equivalence between resonance structures. Although amicable numbers were observed and applied by various mathematicians such as René Descartes and Leonhard Euler during the seventeenth and eighteenth centuries, such subtle number sets have been overlooked and forgotten.

The relationship between God, immortality, and our DNA, for instance, can be seen by using the smallest pair of amicable numbers, 220 and 284. The genuine divisors of 220 are 1, 2, 4, 5, 10, 11, 20, 22, 44, 55, and 110, the sum of which is 284; the divisors of 284 are 1, 2, 4, 71, and 142, the sum of which is 220. Yes, number amicability is somewhat palindromic. The gematria for the word ολον or "whole" in the Hellenic alpha-numerology is 220; the gematria for the word θεος meaning "God" is 284. The gematria for the word αγιος meaning "holy" is also 284. Hence, "God" is universally ubiquitous, *always* existing in and of the whole Universe at the same time. That's what the numbers told Pythagoras, what both he and they tell us.

According to the Bible, humans were created in the image of God, immortally: "So God created mankind in his own image, in the image of God he created them; male and female he created them."[4] The translation of immortality (*immortalis* in Latin) is "deathlessness," "imperishable," "endless life;" *image* comes from the Latin word *imitari*, "to copy," "portray," "imitate." The relationship between the image [of God] and immortality may suggest that human beings significantly embody the energetic and spiritual essences that constitute a person's soul or inner being, reflecting the individual's profound connection to the divine and eternal. The phrase ο ανθρωπος ηταν αθανατος εα or "man was immortal, let it be" is equivalent to the gematria of the phrase η χρυσή ιερή φυλή or "the golden holy race." Both phrases share the same gematrical value 2377 and harmoniously resonating motifs.

The fact that humans were intrinsically immortal then but aren't today is directly related to an evolving construct of our mental and physical states that allows people to prevent aging and death, perhaps physically, but certainly metaphorically and aetherically. Therefore, archetypal human DNA is quite different from what scientists believe it is today. This is the genetic material I identify as 12-stranded DNA. It is the pathway to our immortality, as we will see in the next chapter on how our memories, both current and generational, synergize with our DNA for our healing and evolution.

The Art of Memory and Generational Healing

We shouldn't treat art, literature, and music simply as entertainment. In the real world, they educate us about emotions and empathy so that we can remember our deepest desires and interconnectedness.

The way we can achieve bliss is not through buying more stuff, adding more stamps in our passport book, or drinking with company, but by unraveling who we *really* are. Not only is honest memory the basis of my approach for discovering our 12-stranded DNA, but it is the foundation for reclaiming our life, passion, and purpose.

To begin, we need to understand that true healing doesn't begin with our present circumstance, but we must reach backward, healing trauma we carry forward in order to effect *generational healing*. Generational healing is a continual cycle of remembering and activating memory. Here we will explore how memory underlies both our mortal and immortal nature, and how to both activate and integrate these practices while honoring our place within nature. Memory isn't separate from our DNA or brain because *they* are memories. Memories are a faculty of universal consciousness. Without memory, one cannot be conscious. Memory is not only the storehouse of the mind and body, but also a critical constituent of the individual and collective psyche informed by our awareness of internal and external realms of exis-

20

tence. Within the given omnipresent consciousness, my focus is to help you awaken your memory and empower thought processes that you've always tapped into, and will continue to, regardless of form.

MEMORY
Dance of Brain and Genes

The brain regulates our human experience by translating motor sensory information into coded information and converting genetic memory into function. Conventional theories treat memory as a function of the brain. However, memory is not centralized in the brain; it is predominantly in our genetic material. The brain senses what it needs to activate within our DNA—that's how creatures are so smart so soon after birth—and hence the brain is more of a tuning system to direct and retrieve information from DNA and other planes rather than being the source of the information. The turtle wakes up and its genes tell it to head for the nearest water. The cub knows how to prowl like her ancestors. The octopus has nine independent mini-brains, one located in the head, and each arm contains a brain. Two-thirds of an octopus's memory is spread throughout its arms. Conceptually, the composition of human memory is spread throughout our entire body like that of the octopus.

There are many experiments showing that memory cannot be localized to specific parts of the brain. The neurophysiologist Karl Lashley trained rats to do tricks. He removed bits of their brains—sorry, rats, that's how our neurobiology research works—to examine whether they would still do the tricks. After taking out fifty percent of their brain mass, he was amazed to discover that the elision had no effect on their learning or memory.[1] Another scientist, Wilder Penfield, concluded in his book *The Mystery of the Mind* that memory was not stored inside the cortex.[2] In his studies, he explains that abstract thoughts evidently do not come from the brain. Stimulating the temporal lobes of the brains of epileptic patients, he found that the impulses mobilized past memories, but he still couldn't indicate memory forms in the brain. He was stimulating the cortex likely as a circuit to activate genetic memory.

Nor is memory limited to synaptic plasticity. Bacteria exist without a nervous system and yet display primitive memories. Nearly sixty years ago, Julius Adler used *Escherichia coli* to explore basic biochemistry and molecular mechanisms of learning and memory underlying sensory-motor regulation. He showed that *E. coli* was capable of comparing past and present events and then using negative feedback—showing memory and learning—to decide whether it would stay on course or switch direction.[3]

Memory is an embedded intelligence that encompasses metabolism, growth, kinesthesia, and heredity in a biological system. Beyond its resilience, memory weaves and joins imagination and DNA, unconscious potential and conscious awareness, and spirit or aetheric DNA and corporeal DNA *dually*. Memory is the soul of DNA as DNA is the mortal *akasha* of memory, our cord for life and our strand of universal possibility and learning.* Our cognitive experiences would be meaningless, aimless, and inert without deep components of genetic memory.

Memory and learning come from the summation of cellular interactions. Memory is adaptable, an interaction between organism and environment from which identity—whether worm, bacterium, or mammal—arises. Memory isn't a formless, emotionless map. It is a heart intelligence that encapsulates the soul's imagination. Souls contain deep memories of creation and unity—that's what they *are*—just as much as physical memories correlate with our sensory and cognitive experiences. Emotions influence our focus, perceptions, reasoning, problem solving, and memories. Awareness of our emotions in our day-to-day lives informs our learning experiences and helps retain memories. Emotions are a multimodal entity that defines our empathic relationships and interprets the meaning of the world. Together, this heart intelligence and emotive firing of memory allows us to move beyond our present state and channel the grief and trauma of past generations.

---

*In classical Sanskrit, *akasha* preserves its overall meaning "sky" in Eastern traditions and languages. The term is shared among different religions and therefore the word's philosophical and intended meaning will vary. The word's intended application in this book is aligned with its corresponding translation as "aether," a principle and elemental basis that this text extensively references. In this context, *akasha* is the innermost being of a person or capacity.

GENERATIONAL HEALING

Generational healing is dependent on our becoming an ally with our grief, transcending our current root beliefs, thoughts, and feelings to activate and *manifest* memories. Australian Aborigines embody this concept during long walkabouts—treks routed along sites with inherent power or esoteric energies, called *songlines*. Generational healing begins with setting boundaries according to the decisions of our higher selves; hence, resting, setting limits, intentional breathing, and celebrating grief itself may seem like a new take on healing, but it is the oldest wisdom of our ancestors.

Grief sits on a high throne, having authority in *each* way it affects our transformation and growth. Ordinarily, people think we can heal grief, but grief can't be treated because it serves as a bridge between mortal life and immortality. You don't heal grief; you metabolize and transform it.

It is important not to confuse the original essence of our generational healing, which is to seek our truest authentic selves. Its flame, guidance, and remembrance *are* perpetuity. We continually confront our real selves, not to depend on a scheme or prop for the remainder of our lives, but strive to evolve authentically with knowledge and wisdom. Above all else, we lead with a sense of passion and hope that the vibration of these healing memories will be remembered in the next generation. Generational healing is our immortality.

Let's look more closely at immortality as a function of our memory and a focal point of generational healing.

IMMORTALITY

Close your eyes, and think about the word *immortal*. What visual image or feeling does it stimulate? If nothing comes to mind, explore other sensations that your inner intelligence offers such as sounds, tastes, smells. Then explore the origin of that thought. Even if it's recycled information, it's still an inspiration because you resonated with it and brought it into yourself as a different form of you and it. It is now a part of your being. You *are* immortal.

So often when we think about immortality, we think about living forever in the same body. That would be a nightmare, and it is fortunately not the case. Our thoughts, visions, movements, and memories are, in a sense, holographic emitters of immortality. People fear mortality because of its distinction between one life and the next. But the distinction is only a manifestation of memory. Memory is a cosmic language. Once a person has an experience, we are starting to understand that it is recorded and archived in DNA for the rest of that person's life and can never be lost. Beyond inheriting physical characteristics from our parents such as eye color or blood type, we know that traumatic experiences can leave a chemical mark on a person's genes, and result in their being passed down to future generations epigenetically. But I am approaching memory and where it is marked as an indicator of immortality through something more vast and important. Memories are classified as an estate of the Mind. The Mind being the center of intelligence of not only our body but the center of the Universe. The Mind is a faculty of consciousness and thought that is not exclusive to the brain or the dynamic energy that vitalizes DNA. The Mind, as the fullest expression of human existence and our connection to the galactic realm, collects information in terms of thoughts, imagination, will, remembering, and sensing as it receives and absorbs frequencies that we cannot calculate but feel and experience. How do I know this? Because I cannot translate why or how I experience precognition or am led to information outside of conventional algorithms. It led me to writing this book and discovering how certain sophisticated sound frequencies inspired by the geometry and gematria of our DNA impact water in the way they do, which we will see in chapter 7. I would not have discovered that if I hadn't listened to the calling or message. It may be intuition, but the Mind won't cooperate with new realms if there isn't the will to do so. The Mind is the keeper of our immortality, and only when we choose to explore this concept do we begin to advance what meaning itself is. Immortality is the moment at which memory becomes defined and unconsciousness ceases to stream in willy-nilly sparks and waves. Your unconscious being, like your subtle body and proto-DNA code, is immortal.

The Latin word *manifestus* means "apparent," "clear," or "evident."

The ability to remember is a manifestation—a clearing—that occurs when a memory is activated. The material world distorts the meaning of manifestation by depicting incoming material. It creates a mask, concealing its real definition. Yet memory wants to be revealed—not rejected or blocked out—because it may have traumatic feelings attached to it. We must try to evolve our stance toward our most apprehensive, uncomfortable, and surprising memories with creativity, so we can be assured that when we allow a memory to be revealed, it incarnates to a new beginning, even when we feel perplexed by the mystery, doom, or shame of something that we have either experienced in our lives or can't assume nor understand. Instead, we tend to resonate only with its ending, leading many of us to reject memories and delay the essence that wants to be revealed. However, when approached with mindfulness, embracing memories can lead to a state of harmony and understanding.. The seamless and organic transition between the incarnation of a beginning through the end is hope. Mortality wants to be revealed. We can't mask memory, the "who" inside of us longing to be allowed and remembered. Memory is where our collective reverence is reaching toward but has not yet reached our ego self.

Within its own purposeful intention, each being's unfolded memory *longs to be reached*. It longs to know who and what it is. It is our destiny to remember. Without the reactivation of our memory, we cannot reach new dimensions and understandings.

Healing through Memory

Here is my story of emergence by activating my memory storage of my family's lineage while racing in Australia.

It was my first Ironman triathlon, in Port Macquarie, Australia, in 2017. Without formal knowledge of swimming or owning a bike or running shoes of my own, I signed up. It had been nearly seven years since I played tennis—once my forte—or done any type of athletic activity, but I wanted to honor my dad who, in 2016, passed away seven days after he and I did the Lake Monticelloman triathlon together. His sudden, unforeseeable heart attack left me devastated, and I decided to remember and honor his life by racing on my own in Australia.

Within the course of the year, I worked with a triathlon and swimming coach, rapidly advancing my skills and preparing my body and mind. Learning about triathlons through negative feedback mechanisms within a short period of time showed the same molecular-cellular adaptive qualities as bacteria. Through bike accidents, ice-water swims, and pangs of grief, I ran. I changed my diet to highlight fresh smoothies, juice, and generally healthy eats, and spent most of my time outdoors in the sun, rain, or hail, frigid or sweaty. My morning wakeup call was 5:00 a.m., followed by evening training after work. Sleeping, working, and triathlon initiation became my life trio. I dedicated the year to healing through triathlon energy, and I became a happier person.

When it was time to race in Australia, I was ready. I went alone to be fully present for the track and my dad's spirit presence. My training mates told me that race day is the fastest day of your life, and they were right. The course began with a sunrise loop swim in the saltwater Hastings River, followed by a bike ride that toured dreamlike Pacific beaches. The high-energy run was surrounded by thousands of cheering volunteers, supporters, and locals dressed in comical costumes to send us energy. My training team and family cheered me on from afar as the event streamed live.

At multiple points in the race, I experienced activated memories of my dad that came in different forms of quantum energy, meaning, and inspiration. They were "activated" because I didn't just think about him, he communicated with me through the energy of the race. I could hear his voice talking with me on long roads as I cycled or ran. I felt his warmth. I sensed guidance from his spirit not only on the track, but in my life. It was exactly what I had hoped for.

THE INTERCONNECTION OF
ALL LIVING FORMS

When activating our memories, practicing generational healing, and acknowledging our immortality, it is extremely important to honor our place within nature. To do this I propose a two-part moral code—an empathic moral model. The first part is to recognize that not only are all living forms interconnected, but also the basis of life for all diversified species, whether vertebrates or invertebrates, is equal despite their level, nature, or intensity of emotional intelligence. In the pursuit of cultivating compassion for nature, human beings can activate their mental disposition to consciously evolve with nonhuman entities. This is a critical part of the future of the mind, memory, and generational healing as it encompasses a sacred reverence for *all* life.

Charles Darwin once said, "Sympathy beyond the confines of man, that is, humanity to the lower animals, seems to be one of the latest moral acquisitions. . . . This virtue, one of the noblest with which humankind is endowed, seems to rise incidentally from our sympathies becoming more tender and more widely diffused, until they are extended to all sentient beings."[4]

"Lower animals" can also manifest similar emotions to humans, including happiness, sadness, pain, playfulness, terror, courage, and suspicion. We see this behavior with puppies and kittens. I watched insects behave as though they were playing tag with each other, and squirrels screech and rattle at one another when boundaries are violated. I've watched the curious antics of chipmunks along my path while I hiked the Appalachian Trail. My cat Bijou shreds a bag of bread without consuming it, to show me she is displeased when I leave her home alone when I travel. When I visited a farm in Blackstone, Virginia to research animal frequency, I met young Muffin, a baby goat that was born with a deformed shoulder. She walked with a limp. Yet, when she heard the call from her owner, she ran with great enthusiasm, and one would have never guessed she had a disability. Little Muffin is aroused by the connectedness that each mammalian being intrinsically longs for, in and of itself and with its surroundings.

Empathic perceptions that we feel for a member of a given species are scientifically identified as anthropomorphic projections, which are attributes of human traits, emotions, or intentions given to nonhuman entities.[5] Anthropomorphism derived from the Greek *ánthrōpos* meaning "human" and *morphē* meaning "form." Although traditional scientific models resist endowing animals with human emotions, animals and humans get excited by the same emotions.

The second component of my empathic moral model relates to

the dichotomy between mortality and immortality. As I mentioned earlier, mortality wants to be revealed, and immortality is the unconscious being, like the subtle body. Ironically, humans fear mortality but act immortal through a physical design, pretending their incorporeal existence does not exist. When humans are confronted by death through sickness, suicide ideation, starvation, or an unforeseeable act of nature and survives, their sensitivity and perception of the world commonly shifts. The memory of the "near-death experience" stimulates the individual's course and tracking of a person's purpose. We come from death and we go toward death, as much as death comes from life, it will go toward life, as Socrates said. Thereby, should people choose to accept their mortality as their "guide" in manifesting their purpose in the world despite their conviction of the afterlife, they are consciously embodying their soul's evolution and ascending—whether or not they have experienced a glimpse of death. Such a fundamental beginning raises not only the vibration of the soul, but also the soul's relationships' increasing empathy and morale, and lifting the frequency of the planet.

Once we have grasped our immortality and embraced the empathic moral model, we can begin the work of memory healing, starting with three practices and incorporating the power of sound. Sound and vibration are critically important tools that we will explore in great detail in chapters 8 and 9. Here we see how it underlies memory activation.

SOUND AND MEMORY ACTIVATION PRACTICES

Sound activates unconscious emotions, linking to all memories, including past-life memories. So, deeper memory is what connects us to the wisdom of our primordial ancestors. You know this! Have you ever experienced emotional arousal at an opera? Have you noticed how a nostalgic singing voice may spring a surprising poignancy within you? Maybe when you sing, it causes a burst of tears or a smile? Pick a song or melody, maybe a lullaby. Try it. You may not understand why your body is responding in the way it is, especially when the tone isn't intrinsically sad or bothersome, but the music connects you with an ancient element within. That element feels like your original home, a place that is safe and where you belong.

The uniquely Welsh word *hiraeth* means longing, a pull on the heart, or a type of homesickness. When we travel and see new places, visit our childhood home and revisit early memories, gather family or friends during holidays, or find ourselves in recreational environments such as bars, retail stores, the movies, or athletic games, you may experience diffuse hiraeth. It can feel dreamlike, nostalgic, evocative; it may be acknowledging, fulfilling, or exuding contentment. We are all on a similar track in feeling warmth in the sense of hiraeth.

Beyond hiraeth, one of the oldest forms of therapeutic sound is the straightforward repeated mantra.

Mantras

Mantras are simple chants, spoken aloud or silently, of words that are packed with energy and influence. The present world calls it psychology and adds medical science to it, but mantras have been around for thousands of years. The authors of the Vedas understood the power of the word, and the Hindus passed down their sacred mantras generationally. Buddhists may chant thousands of times a day. Muslims repeat the Qur'an for hours upon hours each day. Jews chant from the Torah and store it in a sacred chamber of the temple. I recall the chants from my bat mitzvah nearly twenty years ago! Priests and Catholic mystics also chant and repeat prayers, such as the Rosary and the Liturgy of the Hours among others. It is no surprise that the secret is in the movement of the sound repetition in the words that takes place on cellular and vibrational levels. A word beyond the surface is a thought. A thought is a word, and equally so to our feelings, voice, and surroundings. A word isn't always spoken; a word is heard and felt by the soul.

Chanting enhances memory, entrains the brain, stabilizes breath, transcends the quotidian mind and induces altered states of consciousness with their innate wisdom. A study combined advanced multimodal electrophysiological measurements and neuroimaging to investigate the neural correlates of chanting Amitabha Buddha.[6] Evidently, increased endogenous neural oscillations occurred in the low-frequency delta band in the posterior cingulate cortex. The strong delta waves localized in several brain regions during slow-wave sleep have been proposed to act as an inhibitory brain oscillation that prevents interfering or distracting frequencies from disrupting internally focused concentration.[7] As a result, religious chanting is an active practice that leads to a self-modulated state of entrainment.

Yet, the brain is not the only complex organ that commands task-evoked responses, such as thinking, emotions, communication, and memory. Sound frequency, as a form of vibration, is also absorbed through the body's largest organ—skin. The utility of entrainment caused by oscillating vibration plays an important role in evolution. For example, the oscillation of cyclin-dependent kinase drives other periodic events, such as DNA replication and chromosome separation, during the cell cycle.[8] The easiest way to visualize directionality of entrainment is to tap a water glass with a pencil or spoon and watch the water vibrate and create sound waves through the water. A part of the audible sound is spent on the reorganization of the water's molecular structure. In the same way that our skin's epidermis contains seventy percent water, sound frequency plays a profound role on yielding a stable entrainment structure that inherently impacts the capability of our DNA. DNA is the basic vehicle of entrainment, mediating evolution, and mantras are a way of remembering to activate it.

Reflect on a mantra that feels inviting to an ancestor or to your living family. It can be as simple as, "I'm here," or "I remember you." Use it as long as it is active. When it loses its energy or fizz, then create a new one. Whenever you're in a private space, try your mantra and observe how you feel when you say it. Do you sense a deeper connection to the moment and to the flow of time? What do you start to visualize as you say the mantra? How does your body start to feel?

Imagination

To fully unlock the power of these meditations, we must engage and exercise our imagination. Imagination is an intrinsic element of our soul's desires and expression. In Sufi tradition, imagination is an uncontrolled thought, and the light used in guided visualizations is a tool to self-direct, organize, and control the imaginal plane. Light can mean symbolism, metaphors, or characters to awaken your conscious mind and actively participate with your imaginal energy system. Not only do I walk you through the meditative cosmic breathwork practice in chapter 10, but also guide your imagination to prompt your thoughts into programming your body's experience with your breath. In the meditation on composting grief in chapter 12, I cue your thoughts to imagine a channel for your emotions to flow down to the Earth's soil and exchange your body's information with the soil's microbes. Our imagination is the inner artist, and she begs to be expressed as we create our own canvas depicted as reality and its destination, a memory.

Harnessing the power of your imagination, try this meditation to activate your inner child. The inner child is our earliest memory to what we associate as an ideal world, which ties into the science, philosophy, and mysticism of the number 12 in the following ways: The Greek word for *child* is παιδί or *paidí*, and its alphanumeric gematria is 105. One of Plato's most influential contributions in the way we think about the world was introduced in one of his dialogues, *The Republic*. In this text, he asserts his Theory of Forms (also known as Realm of Ideals) and the concept of an ideal world. The phrase "the ideal world" in Greek—*o ideato kosmo* is 1260. If we take 105 and multiply it by 12 (105 × 12), we get 1260. The numerical relevance is inevitably and mathematically interconnected. To add another layer, Plato's name, Platon, is only one digit higher, 1261. Plato believed that names are given not by coincidence or accident, but follow some sort of mathematical algorithm, which is still an unknown concept in human history. It is a type of formalism. The intersection at which the integral geometry of ideas (otherwise the nature to these numerical proofs) meet presents a possible approach to navigating our modern day lives and spirituality with mindfulness and a higher level of consciousness. Additionally, this approach emanates Plato's teachings of the ideal world in that the archetypal forms are "abstract, perfect, unchanging concepts or ideals that transcend time and space."

Inner Child Practice

Close your eyes and imagine you are playing outside, surrounded by sunlight. Make the shape of a balloon in your stomach by taking a deep breath. This is your inhale. Imagine the balloon has a tiny hole the size of a needle in it. It slowly deflates. That is your exhale. Take one more inhale and exhale.

With your eyes closed, imagine you are in a forest, surrounded by tall trees. The soil feels soft and mushy under your feet. Feel the bottom of your feet sink into the soil. Don't be scared, it only wants to get to know you and be your friend. You can move around freely and anywhere you like. A special animal wants to come out and greet you. Are you ready to see who they are?

Out of the trees comes a shining white horse, slowly walking up to you. It wants you to know its name is Star. It invites you to gently pet the top of its muzzle. Star wants to play a game with you. Are you ready to play? It's called, "Where's the butterfly?"

Star says to you, "Imagine a wave of light starts from the top of your head and moves slowly down to your feet. That is the butterfly. It is similar to feeling a breeze from a wind, warmth from the sun, excitement in your belly, or your heart dancing with joy. Any movement that comes up is a sensation.

Start with the air above your head, spend 10 seconds here. Then go to your head, spend 9 seconds here. Go to your neck; spend 8 seconds here. Go to your heart; spend 7 seconds here. Then to your stomach; spend 6 seconds here. Then to your hips, spend 5 seconds. Then to your legs, spend 4 seconds. Next are your feet, spend 3 seconds. We will end by thinking about the patch of the soil. Spend 2 seconds here. When I say 1, tell me where you felt the butterfly."

(Pause for 5 seconds, then say 1.)

After your child reveals what they feel, continue the visualization: Star is so happy that you found the butterfly! Now it is time to come back. The next time you meet Star, you can play this game again. Taking an inhale and exhale, gently open your eyes and come back to the space.

The exercise may be childlike, but the sensations it evokes are guileless and real.

Soul Memory

The following prompt presents a series of questions that will cue your self-reflection on memory stream and present moment. The goal is to listen and be receptive to what comes up for you. You can decide how to proceed from there at a future time. Practice this prompt consistently and pass it on to future generations as loving guidance. I mean to stimulate your cognition, thinking, and your DNA.

Using a mirror, take a look at yourself, and give yourself a warm greeting, as if you are meeting a dear friend. Ask yourself these questions in the mirror aloud, just as you normally would in a casual conversation. Call yourself to be sincere and know you are safe to respond transparently without anyone watching. Listen to the word of your soul. Take time to answer each before moving on to the next one.

What do you call yourself?
Who are you?
Who am I?
Why?
Why are you here?
What do you remember about yourself?
Where are you going in your life?
Where do you want to go?
How do you want to be seen?
Why?
Tell me about your love.
Is there a part of you that needs to be seen or heard at this moment?
Where do you want to feel more love in your body and life?
What will you do in the future?
What will you do in the past?
What will you do now?

Now that we have sensitized our being and raised our vibration through memory and generational healing, we will bring these tools with us as we explore how to transform our evolution, another preliminary step to fully understanding our 12-stranded DNA.

Devolving to Evolve

Our mind is a spinning bobbin on a wheel. From one thought to the next, it pulses persistently. It forgets how to synchronize with the moment. Mindful moments are a basic human need for processing the outer world in the inner world so that thoughts can interact with both realms in an equal, honest, and authentic fashion.

While we seek metamorphosis, change cannot be done in a single step, through a pill, or prescribed method. There is also no finish line or medal. Instead, we get something more invaluable: we get to evolve. But *before* we evolve, we have to fall back to get ahead. We have to *de*volve.

EMERGING WITH COMPLEXITY SCIENCE

Before we begin to look deeply at DNA, let's look at evolution and a thing called complexity science. Complexity science is how a "large collection of components—locally interacting with each other at small scales—can spontaneously self-organize to exhibit non-trivial global structures and behaviors at larger scales, often without external intervention, central authorities or leaders."[1]

When life emerges, it self-organizes from systemic throughput of energy. It is concomitantly organized by nonequilibrium dynamics, complexity, and chaos. Complexity epitomizes how simple systems interact with one another to express emergence and self-organization. An excellent example of this complexity is the humble slime mold. Being a simple organism with no central nervous system does not mean it is primitive. Slime molds have incredibly sophisticated adaptations to energy, shape, and function:

> When food supply, temperature and humidity are favorable, the slime mold lives as separated, individual cells. But when conditions are strained, these cells coalesce to form a single, larger organism. Time-lapse photography shows that the organism then crawls very slowly along the forest floor in a unified search for food. When its environment improves, the cells separate and become solitary foragers once again.[2]

A slime mold takes simple parts interacting together to achieve goal-oriented behavior without any additional signal, stimulus, or synapsing. Communicating with pheromones, its cells can find other cells that have already found a source of food. It has properties of emergence and self-organization. It shows us a "bottom-up" approach to information as well as a model of the interaction of simple parts in complexity science.

A WORLD OF COMPLEX SYSTEMS

Science tracks emergent systems. With each emergence, a new order branches out of a prior one. Without physics, we wouldn't have chemistry. Without chemistry, we wouldn't have biology. Without biology, we wouldn't have ethology, ecology, or psychology. Without ecology, we wouldn't have anthropology and sociology. Consciousness is a series of emergences within the mind.

In the last forty years, theorists have noted how principles governing complex systems also govern economics. The notion was introduced by Adam Smith in 1776 when he used it to describe how goods and services are distributed efficiently without central government planning.[3] Since then, complexity science has been used to map not only political, socioeconomic, and cultural systems, but also physical and biological systems. Though it ordered everything I was studying in school and in my pharmaceutical career, we behaved as if complexity science didn't exist and we could control nature without complexity. Consider that! It's like thinking that you can walk through a rain shower without getting wet.

Even the structure of civilization can be seen to evolve from reducing its components: land became territory, territory became property, property was given by deed, the deeds are mortgaged, which itself was accessed through money. Money was derived from gold coins extracted from the earth. Science itself has also abstracted matter on multiple levels in order to commoditize health, transportation, communication, defense, economics, agriculture, energy, and conservation.[4]

A great example of a natural process turned into a commodity is the case of Kopi Luwak. Kopi Lowak is an Indonesian coffee bean digested and fermented by the Asian palm civet. This catlike mammal roams the forests at night, eating ripe coffee cherries and excreting beans. The beans are handpicked, cleaned, and roasted to make cat-poop coffee. The reason people seek Kopi Luwak is because the palm civet eats the best, ripest cherries, so you don't end up with inferior, unripe beans. Then, while passing through the civet's stomach, the cherries get stripped of their fruit exterior, or pericarp, and thus evade the growth of mold, resulting in a tastier sip of coffee. A cup of Kopi Luwak coffee costs between $35 and $100, while the per-pound price ranges from $100 to $600.[5] The downside of this process is that palm civets are kept in tiny cages on coffee plantations.

We see complexity in my example—a combination of nature, technology, and culture—leading to a product and an ethical dilemma. The synergy of the civet's diet and the coffee plant's evolution intersect, with human acquisition of wealth as their driving force. We must be mindful that increased consciousness does not always mean increased harmony or a better way of living. In modernity, collective consciousness has taken a high toll in cobbling together a zombielike condition of survival and wealth, cut off from an internal world. Much of the human population has accepted this dissociation as fate—accepting the feeling of being fatigued, depressed, and hopeless. Behind this physical world's dilemma sits a subtle plane of elements, forces, and karmic energies that are waiting to be unraveled; we are unconsciously seeking a forgotten memory.

This is where the practices I cultivate and teach find their place. I present a separate frame of devolution that appears as the "non-progress" side of the story in the face of evolution. It involves changing our expectations of what evolution is supposed to look like to one that embodies our thought development into physical applications or rituals.

DEVOLUTION

Devolution requires a shift in intelligence toward an entity, an independent artifact containing information of life. An entity can be anything: your phone, job, furry baby, health freedom, movement practice, or even a thought. It is the sun or moon, an animal, a sound, a dream, a memory. It is a medicine that you use for your mind, body, and soul. An entity is also your perspective of yourself. The space in between is the way we relate, function, and evolve with the entity. Our society is conditioned to resonate to our relationships with entities as either toxic or positive. Such dualism creates unending methods of detoxing, decontaminating, depolluting, and decrystallizing without purpose or intention. It *defers* our power with these entities that are meant to shape our evolution, growth, and potential.

One example of a shift in intelligence toward an entity is behavior with a virus. Viruses play an important role in how we reinvent ourselves. Although the world has villainized viruses for centuries, they can be medicinal; viruses carry knowledge. The human body contains millions of viruses, and their abundance in us is not by any shortcoming or mistake. Endogenous retroviruses have been helpful in forming biological memory, developing a placenta, and the evolution of most genes over time.

Sequences from viral genes code for proteins in our cells from generation to generation. The units of DNA, like the sounds of language, can be put together in different ways. An allele that expresses itself one way in a virus may express itself another way in a worm, another way in a tomato, and another way in a hedgehog or in us. Splicing DNA is not just about making phosphorescent or frost-resistant tomatoes or pest-resistant corn; it's about making whole phyla and kingdoms of life.

Sovereign ensoulment is applying intent, action, and belief in an entity to activate memory. But we have to begin somewhere on a conscious plane. Creating subtle shifts in your daily regimen is a place where we can start. We can begin by having an internal dialogue with our own DNA.

EDITING OUR DNA

DNA and RNA editing may be a significant contributor to evolution. Editing our DNA is not in our present repertoire, but it is in nature. Species of octopuses, squid, and cuttlefish species, "routinely edit their RNA sequences to adapt to their environment."[6]

In Chapter 1, the composition of DNA was discussed. DNA and RNA (ribonucleic acid) are composed of nucleotides, consisting of a nitrogenous base, a ribose sugar molecule, and a phosphate group. RNA is a polymeric molecule (poly-mer meaning many-parts) and refers to one of the most important biological macromolecules in addition to DNA, proteins, lipids, and carbohydrates. RNA is composed of four nitrogenous bases: uracil (U), guanine (G), adenine (A), cytosine (C) and known to exist in a single-stranded form in nature. There are, however, unique RNA viruses that are double-stranded. RNA is formed from DNA, which contains the master instruction set through a process called transcription within the nucleus, and RNA creates proteins through translation within the cytoplasm. DNA stays in the cell's nucleus, but RNA is small enough to travel outside the cell's nucleus to the cytoplasm to make proteins.

Protein synthesis involves three types of RNA: messenger RNA (mRNA), transfer RNA (tRNA), and ribosomal RNA (rRNA).* Small but mighty, mRNA contains the instructions or "programs" the cells to create proteins naturally. The purpose of tRNA is to bring amino acids to the specific sites in the ribosome to create a protein. The emphasis of the tRNA is in its hairpin loop structure: one side of the tRNA contains the anticodon sequence and the opposite side is where the amino acid is attached (such as methionine, serine, histidine, leucine, tyrosine, etc.). Using its anticodon, which contains a sequence of three nucleotides that code for an amino acid, it recognizes and decodes its corresponding mRNA codon. The anticodon and codon base pairing is not complex—think of it as a system of classification. Imagine going to an extravagant, wholesome library such as the New York Public Library to search for a specific book. Most libraries use the Dewey Decimal System, a classification system used to organize

---

*The purpose of rRNA is to assemble the structure of ribosomes in the nucleolus.

books sequentially, assigning the first three digits to its corresponding discipline (e.g., generalities, philosophy, religion, social sciences, languages, and on). Knowing the beginning of the first three digits specifically, the reader can precisely locate the book and check it out of the library. This is how tRNA recognizes the appropriate codon/anticodon base pairing. It brings arriving amino acids, recognizes and decodes the mRNA sequence by codon. When a stabilized codon and anticodon pairing forms a peptide bond between amino acids of the tRNA, the tRNA "checks out" or moves onto the next section of the mRNA sequence on the ribosome until a stop codon is reached. Voilà! Protein delight.

> When such an edit happens, it changes how the proteins work, allowing the organism to fine-tune its genetic information naturally without actually undergoing any genetic mutations. . . . Researchers discovered that the common squid has edited more than 60 percent of the RNA in its nervous system. Those edits essentially changed its brain physiology, presumably to adapt to various temperature conditions in the ocean.
>
> DNA sequences were thought to exactly correspond with the sequence of amino acids in the resulting protein. However, it is now known that processes called RNA editing can change the nucleotide sequence of the mRNA molecules after they have been transcribed from the DNA. . . . In some—but not all—cases, this event will change, or "recode," the amino acid encoded by this stretch of mRNA, which may change how the protein behaves. This ability to create a range of proteins from a single DNA sequence could help organisms to evolve new traits.[7]

Gene editing is particularly complex to the mind. Several studies explored meditation-based interventions influencing gene expression toward wellness and have identified gene-expression profiles that *are* sensitive to such practices.[7] Such interventions were found to downregulate epigenetic pathways related with inflammation, cell aging, and depression.

In my own practice and research, I developed a simulation for 12-stranded DNA activation by combining a guided visual meditation and the healing sounds of handpan music for eight participants. This exclusive meditation emphasized each person's innate ability to activate one's DNA through imagination, emotions, thoughts, and feelings. I believe that by siding with these forms of reflection using the visuals and feelings that I verbally deconstructed, the participants self-reported a subtle, cellular stimulation that supported feeling a biotic reset and overall restoration. I relied on the participant's verbal and written feedback to assess how their body and mind responded. The aim was to observe the participants' perception following the cellular-based meditation, which would reflect in the state of the nervous system that controls blood chemistry.

Controlling blood chemistry enables us to rewrite the genetic algorithms in the nucleus of the cell. One participant reported reduced inflammation in her shoulder. A second participant wrote, "I feel as if everything is in slow motion so I can see what I need to see and then react accordingly." Another participant noted that her inner chatter stopped, and she could experience "thoughtlessness" for a temporary period, feeling a sense of calm in the body. What did the participants do to have these experiences? In this collaboration, I trust each person's effort, belief, and willingness to activate the ethereal content and high dimensionality of our genetic makeup. That is, listening to my voice and storytelling, which weaves a tapestry of experiences, observations, actions, and realizations, and each person's commitment to reflect on this meditation results in a felt frequency shift. Moreover, my hope with this exercise was to present an experience that illuminated each person's autonomy in taking charge of their own frequency and evolution.

Despite the novelty of this field of study, there are numerous publications that illustrate the positive outcomes of meditation-based techniques and systems. Epigenetic mechanisms such as genetic silencing, methylation or demethylation,* and variance are highlighted as general outcomes in such an intensive mindfulness practice.[8]

However, I believe the present gulf between science and spiritual practices deters our emotional and psychological advancement. As a sci-

---

*Demethylation is the removal of a methyl group (-CH₃) and methylation is the addition of a methyl group to DNA, proteins, or other organic molecules. Both are significant mechanisms in gene expression and epigenetic regulation.

entist, I agree we are still far from identifying key genetic markers and understanding the mechanical details of these meditative techniques. As a healer and a musician, I propose what a scientist in industry would not—that the answers to many of our environmental crises may lie not only in our noncoding DNA, the secret tablet holding the completion of human evolution, but also with the self's relationship and active embodiment of love. It is our recognition and desire to reemerge with the frequency of love energy that will activate our mind, and thus our 12-stranded DNA.

∽o∾

Over a year after racing in Australia, I found myself at the starting line of the Ironman race in Lake Placid, New York, once again without any dedicated training. Opening and running a food truck consumed my year, leaving little time for focused triathlon preparation. Lake Placid is not only a famous Olympic center, it's the place with the most sought-after triathlon around because of its legendary course. I wanted this one mightily. I wanted to experience a different kind of evolution. This time was going to be about emotion, empathy, and soul, as much as about cognition, technique, strategy, and overcoming mental hardships. The natural challenges of life and terrain mirrored each other.

The race was complex and, also, nature took its course. During the 2.4-mile swim, my wetsuit wasn't adjusted appropriately, creating a wide gash in the side of my neck. I didn't notice until a volunteer asked if I needed aid to stop the blood running down my back. While they patched my neck, I closed my eyes and listened to the rumblings of the sky. A storm was coming. The officials left it to each athlete to decide whether they wanted to forfeit or ride. "How ya travelin'?" the volunteer asked as she was wrapping me up. "I'm traveling," I said. I took a chance.

The opening climb is always manageable, but the key is conserving one's energy because the course is a two-loop repeat, so I did it twice. Hail bounced on the ground as I reached the apex of the climb. Ironically, this is where the Keene descent began, the most aggressive, steepest hill. "Cheer, do not fear, the Keene descent" is the race-event's tagline. This is where a cyclist can reach up to fifty miles per hour without pedaling or breaking. But I wasn't going to turn around. I saw a lightning crack over the valley, marking the point of my destiny. I thought, What a life this is!

As I started to roll down, heavy rains, hail, and gusty winds challenged my wind-battling aerodynamic bike. Every fiber of my being concentrated on holding the chassis on wheels and looking out for dips and cracks in the road while trying to make sense of what was resolving beyond my wet sunglasses. I felt myself wanting to cry from despair. It was as if this was suddenly the only place in the universe. In that moment a crack of lightning lit the sky, and I remembered. I can't even say what. It wasn't a memory of something. It was memory as memory. It activated and helped me realize why I was here. As I descended past a sign that said, Trucks Use Low Gear, I remembered this journey is my evolution. "I need this fear, and I'm supposed to be here." Later I thought that this course was a classic microarray experiment to help me detect soul "data" about my higher dimensional self. The only way I could extract the information was by putting my physical body through a trial of nature, terrain, and trust.

After eight terrifying miles of descending, I found myself in Keene Valley. It was in the mid-40 degrees Fahrenheit, and I was wet. My toes and fingers were starting to throb and ache. I promised myself that I wouldn't experience hypothermia as I was passing by other cyclists who were already receiving aid on the side of the road. One way of avoiding hypothermia is by pedaling harder and faster to get warmed up, but I didn't find that any different from being guided by broken emotions to absolve stress. Instead, the cold conditions led me to create my own warmth from within, a power I needed in life outside the race. It gave me a chance to experience stillness as restoration. I was no longer thinking about deadlines, emotions, bills, my job, or any stressor. I was simply surviving. The cold woke and saved me. I thought about the things that melt my heart: my family, my ancestors, creation. I loved the universe and my identity deeply in that moment. Love was my salvation, and it activated my memory, my body-mind.

After some time, the sun started to peek through the clouds. The second loop proved much more manageable than the first, and the course *started to feel easier for the remainder of the 112 miles. As I returned to transition for the run, I realized I forgot to pack one of the most critical pieces of running gear: my socks. I took a deep breath and sighed. "Here I go," I thought, and began my 26.2-mile run without socks.*

The marathon demonstrated my mental, spiritual, and physical abilities to myself, separate from the swim and cycling portions. The running terrain was composed of ongoing hills. The sun started to set about seven miles into my run, dropping the temperature rapidly. My engorged feet had seen better days, and I was uncertain if my skin was even attached to the bottom layer of my feet since I felt it gliding in my shoes. I didn't know how to step without feeling pain, so I decided to go for mind over matter. It was a call, something like, "I'm alive, and I am grateful to be." I was able to transcend my suffering because of this thought and started to move differently. I wanted to catch up to the next person just to cheer them on.

The soul tag game that I created became fun, distracting, and I encouraged other runners. My body forgot that it was experiencing pain in terms of suffering. Instead, I embraced pain as information. I sensed the angelic frequency activating me because my feet felt light by remembering to uplift other people! Knowing that they could be supported by another participating athlete was valuable to them and to me. If I could do it, so could they. By shifting my mindset, I created hope for myself and others. In the end, I crossed the finish line in under sixteen hours, just before the cutoff time.

The point in sharing both personal accounts of my race experiences is to exemplify memory activation in two separate modes. The first story summarizes the act of emergence by activating my memory storage of my family's lineage while racing in Australia. In the second account, I reemerged from embarking on a sole mission to support the athletes trekking the same journey based on empathy. The analogue developed: emergence to sensory-motor; then reemergence to "soul-motor." The space in between comprises the emotional transitions that lead to generational memory healing. This is the "progress" side of evolution.

REEMERGING IN FUTURO WITH LOVE

Love is electrical. Love isn't limited to a field of sensation or learning or accompanied by a list of post-nominal letters. It isn't dualistic; it *is*. Love doesn't have absolutes; it *does*. Love is what you experience beyond light: radiance, ground luminosity, energy. Even a machine running at the end of a cord is an expression of love energy, because love is a vibration: the single most cohesive force of the Universe and binding glue of everything, in mind or in matter. Love is cosmic, as much as it is terrestrial or cerebral. It unfolds like the sun bursting out selfless from its corona. Even if it doesn't necessarily manifest in a physical form, love extends everywhere endlessly. If you want to see love, look at the blue sky. It is endlessly dense, and yet you can see through it.

Love wouldn't exist without hope. In a metaphorical sense, I can relate hope to an empirical domain such as the gravitational constant. In Newton's law, the gravitational constant measures the force between two masses and its relationship to the product of their masses in inverse proportion to their distance (input and output of unification). In Einstein's field equations of general relativity, the gravitational constant is the link between the geometry of space-time and energy momentum. Different aspects of each law share a similarity with the concept of hope—it is something that is difficult to measure. I believe that the "dense" aspect of hope can be likened to our mRNA—guiding the mind to make a specific decision, or protein, when we experience resistance, emotional hardship, or the gravity of evolution. Hope is a soul language and a memory.

Hope has become an endangered concept, a slipping memory, a forgotten embodiment, a withering morale, perceived as a lost cause, and a questioned holy practice. It spawns everything: positivism, art, music, healing, food, animals, Bob Marley, coffee, t'ai chi, meditation, pharmaceuticals, sex, alcohol; you can fill in the rest. It is the manna that connects. Even nihilism is impossible without hope.

Truth is an expression of beliefs, unconscious as well as conscious, but most importantly unconscious because we are planting seeds—sounds and sigils—in one another's unconscious. When we try to trick or outsmart our own truth, we lie. We think that a lie protects us from rejection, shame, and guilt—from fear. Deception fuels spiritual stagnation, leading to material idolization, overshadowing deeper connections and values.

In more modest engagements of creative visualization, my hope is for you to be patient and honest with yourself chapter by chapter and always return to your belief in yourself, in the affluence of love in creation, and your hope to ascend into your fullest self. Your sincerity, transparency, and intention will guide you in the direction of your truth because there is no trajectory with more gravity. What longs to be activated in your DNA will be activated.

DEVOLVING TOWARD INCREASING COMPLEXITY

With this premise, we now move to the first building blocks of our practice, beginning meditations to unlock our 12-stranded DNA.

Slowing Down

The first step in addressing our evolution is that we must force ourselves to slow down enough to hear a very subtle spiral clock that is not tracking time but rather the link between aetheric and physical time. Our brain is going to adapt as our mind adapts to thoughts, emotions, cognitions, and changing environments.

With resonant programming and fluid mindset, our body finds time and space—cell time and cell space—to make these subtle changes. Lucky for humans, we're settled on a planet where time and space exist. Feel that you can slow down because you're meant to. Many times a day, repeat to yourself, "There is no rush. I am exactly where I am supposed to be."

Creating Space

Another part of our journey is removing ourselves from the static and noise. That doesn't necessarily mean ambient sound; it means the ever-growing volume in the mind created by decades, a lifetime, or generations of cogitation: charged mental action. When thought catches up to us, we experience burnout, confusion, desolation. The hardware—our body-mind—needs a system update by uploading a new software upgrade. The upgrade begins with accounting for the space between each note or sequence. Vibration needs space to breathe, to even vibrate, and so does our DNA.

Vibration is paradoxically instantaneous while having a life span. It is an *emergence* of information that communicates a story or a set of instructions to the next vibration, a bit like a zip file that doesn't have to be opened to be transmitted and integrated. When the brain receives a vibration, induced from within or externally, it subconsciously triggers hormones that either support the body's vitality or cause it to go into stress and imbalance. The way people respond to vibration depends on how they see themselves in the world and their emotional intelligence.

In most systems of meditation, a person organizes and witnesses breath and vibration within. When the body is grounded, a person can effectively work with it to the deepest level. That is to say, the way we think alters our DNA as well as heals our emotions and supports our genes. Stress, fear, sadness, and anger not only damage our bodies but get into the vibrations of our genes. Alternatively, we can genetically and neurologically heal them through guided thought. Try the following practice to begin slowing down and making space.

Slowing Down Practice

To begin our devolution, try asking yourself these two questions: What do you spend your time worshiping, and how do you convince yourself to do it? Whatever your truth is in the present moment, acknowledge it. This is our starting point together. To highlight the importance of a moment of time, let's observe it metaphorically through the lens of clouds. Clouds exist in unique forms and shapes. As we look up at the sky, clouds pass by usually very slowly. If we want, we can meet each individual cloud. They don't sprint across the sky to get to the other side as quickly as possible. They take their time and, when they do move on, we'll never see that special cloud ever again. Our thoughts are the clouds. As we slow down, we sanctify our inner world. Feel the cloud; keep going at its comfortable pace. Soon the soul can breathe, feeling unrushed.

Take a conscious breath. Think about the next questions: What has changed since yesterday? What has changed since an hour ago? What has changed in the last second? What revelations have come to you? How are you feeling?

Gently, take another conscious breath and close your eyes. Let your feet root in the ground. On each exhale, slowly scan your body, starting at your crown or crown chakra or even esoteric chakras above the crown. Psychically smooth or cleanse anything that feels wrong or odd along the way. If you don't know how, imagine it. Imagination is where real magic begins. When you reach your feet, let your breath rise and fall normally. What inspirations did you experience?

This approach honors the known and uncertain realms of evolution: evolution evolves itself. Take a moment. Journal your thoughts. Create some space, perhaps do an activity or task from your day to day routine. When you wish to come back and revisit your experience, pay attention to the subtle shifts in your internal landscaping. What do you notice? What has changed for you? What do you desire to explore or seek now?

᭒᭒

I am not a yogi, a licensed bodyworker, an herbalist, a shaman, a barefoot healer, an alchemist, or a sensei. I am a woman born in Kyiv, brought to the United States as a girl by prescient parents during the rise of Vladimir Putin. My Jewish heritage is important to me because our ancestors imbued us with remembrance of our roots, togetherness, and faith, even if our beliefs aren't tied to orthodox Judaism. I was raised in Virginia and trained as a scientist. I married by chance into an apocalyptic Christian cult because I thought in school, "How perfect, a Moldovian exchange student (him) and a Ukrainian American (me)." I was sheltered, provincial, and vestal. Yet somehow, like a coffee bean, I was digested by the "civets" of my experience and came out of the transformation with the wisdom of negative capability. I realized that I could hear animals and life forms and receive messages from them. I could transmit sounds. I could receive and transmit ciphers and sacred codes. I understood DNA implicitly from hands-on experience. The 12-stranded form wasn't an abstraction; it was a voice from inside as well as outside me.

The Genetic Archetype and the Dodecahedron Form

A gene's original form comes from a place beyond Heaven and Earth. Its mystical construct tells a story about the most sacred meaning of life that parallels other philosophical, theological, cosmological, and psychological narratives. To understand the underlying order of the 12-stranded DNA, we must begin by examining its underlying archetypes. The term *archetype* dates back to 1736, when it appeared in Sir Thomas Browne's *The Garden of Cyrus*: "So that being sterill before, he recived the power of generation from that measure and mansion in the Archetype; and was made conformable unto Binah."[1] *Arché* means "first" and *type* means "model." Carl Jung described archetypes as "the original pattern of which all things of the same type are representations."

Jung's concept of psychological archetypes drew on the *eîdos*, or "ideas," originally developed by Plato. While Jung turned the Freudian unconscious into a collective unconscious of transformative symbols, Plato expounded upon the essences of pure Form—that knowledge is the ability to capture the world of forms with one's mind. This idea is reflective of my own use of archetypal DNA in personal evolution and healing.

48

PLATONIC FORMS AND DNA AWAKENING

Platonic Forms and DNA share ambiguity and contradiction. Whether it is a premonition or truly a chimerical pairing, DNA and Plato share common themes on genetic essentialism: perfect and imperfect, aspatial and atemporal, latent and manifest, soul and self.

In *Phaedo*[2] and *Republic*,[3] Plato introduces the idea of an objective blueprint of perfection, describing the perfect forms as *idéa* or *eîdos, morphē,* and *parádeigma.* Though Plato's physical world is changeable and unreliable, he says, beyond its appearance is a real or permanent world that is reliable. He proposed that each idea, quality, or thought has a form transcending space and time. In my system, love, animals, humans, angels, positivity, hope, evolution, sound, light, numbers, and geometry come in forms.

The nature of Forms cannot be taught but is present in each person's mind because the soul contains the memory of the Forms from before it was born. Thus, knowledge of Forms is a matter of life experience, taking chances, and recall through memory activation. A gene's original essence and form come from a place beyond Heaven and Earth. Our real knowledge of its form must be the memory of our original acquaintance with it. What we seem to learn is *just remembering.*

In the *Republic,* Plato shares the metaphor of people who live their entire lives in a cave to depict the everyday experience of reality. Shadows are projected by objects passing in front of a fire behind them. They spend their time with the shadows on walls. As they step out in the sunlight, their ability to see is analogous to remembering and experiencing a true mental reality, the world of Forms. A person who is accustomed to seeing shadows for his or her entire life associates real physical events and objects with their dim replicas.

THE GEMATRIA OF FORMS

In my exploration of archetypes, I treat a Form as having real knowledge partially rooted from its underlying gematria value, a point of reference and methodology favored by both Pythagoras and Plato. Introduced in chapter 1, the Pythagoreans discovered five solids, known as the Platonic solids, which have the following properties: each of them is constructed by using the same regular polygon, and all their three-dimensional angles are equal. Let's review the five Platonic solids in further detail.

Every Platonic solid, which is also called a regular solid, is constructed by congruent (identical in shape and size), regular (all angles equal and all sides equal) polygonal forms with the same number of faces meeting at each vertex. Five solids meet these criteria: The tetrahedron is a pyramid made by four identical equilateral triangles; the hexahedron is well known to us as a cube; the octahedron is made by two identical pyramids with a common square base; the dodecahedron is constructed by twelve equal regular pentagons; the icosahedron is constructed by twenty equal equilateral triangles.

THE FIVE PLATONIC SOLIDS

	Tetrahedron	Cube	Octahedron	Dodecahedron	Icosahedron
Edges	6	12	12	30	30
Faces	4	6	8	12	20
Vertices	4	8	6	20	12
Element	Fire	Earth	Air	Aether	Water

The five Platonic solids are classified by their number of edges, faces, and vertices, as well as elements. Aether is the source of all the elements.

Each of these 3-D shapes can be related to one of the five elements of universal creation: fire (tetrahedron), earth (cube), air (octahedron), aether (dodecahedron), and water (icosahedron). According to Plato, the dodecahedron symbolizes the mental nature of human beings connected to aetheric and universal fields, and its numerical value is, not surprisingly, 12.*

---

*Upon completing this book, I attended the Science of Consciousness conference in Tucson, Arizona to present a study on sound and synchronicity, where I discovered the works of physicist Anirban Bandyopadhyay. Bandyopadhyay's talk revealed his invention of an artificial brain synthesized from a time crystal conversing with a human brain using a technology that he founded called dodecography (DDG), an electromagnetic

Figure 4.1. The five regular Platonic solids. The faces constitute the sides of exterior panels. The edges are the straight lines that outline each face. The vertices are the points where two or more edges converge. The solids include (1) a tetrahedron, (2) a cube, (3) an octahedron, (4) a dodecahedron, and (5) an icosahedron. This figure is taken from Marrin's *Universal Water*.[4]

Gematria values are knowledge from the Form itself. I can't make up numbers—they already exist—aligned, quantified, and given operational values and constraints. I simply observe their equivalence of Form. I have adapted Plato's 2,300-year-old theory to my system, which we will look at more closely in chapter 5.

THE FORM OF DNA: THE DODECAHEDRON

Nature demonstrates Forms and archetypes through geometry. Geometry differentiates shapes and spatial arrangements: chirality, ratios, crystallography, numbers, ratios, pathways, spins, scalar properties, and phases. Geometric forms offer molecular lattices behind archetypes. For example, the function of tetrahedral molecules in nature

---

(cont'd) advancement of electroencephalogram (EEG). In his talk, Bandyopadhyay says, "If you've known language of nature, you can do wonderful things. And if anybody claims to you, they have understood consciousness or created a model of consciousness or understood what is information in nature, ask them to build something so they can create sand inside the human body and do not get any response from the protein. So we created this machine. And if you send normal drug, if you take even paracetamol, you will find 500, 600, 700 protein up regulation down regulation. You send this one, which we are GML and PPM with our language, we have modified it, we get 40 to 42 maximum or 5 to 6 highly expressed." In his book *Nanobrain*, Bandyopadhyay elaborates this concept by presenting a frequency fractal model of the whole brain, made up of 12 conscious decision-making tensors. His team divided the brain into 12 different bands, representing 12 conscious components, hold 12 types of memories by 12 classes of rhythms.[5]

depends on their vertex angles and chemical constituents. Methane and water, both tetrahedral molecules, differ by polarity and electronegativity because their geometric structure is distinctly different. Even the transition from methane liquid to methane gas comes from the same molecule and depicts a shift of archetype. Chlorophyll and hemoglobin are almost identical in molecular structure but differ in their center ion. Chlorophyll contains magnesium (a green vibration) and carries light for photosynthesis, while hemoglobin (a red vibration) contains iron and carries oxygen to the body.

Nature manifests other forms of lattices in triangles, tetrahedrons, cubes, pentagons, hexagons, octahedrons, dodecahedrons, prisms, and variously shaped polygons that present millions of archetypes and trillions of molecular and biochemical structures and substances. I will focus mainly on the natural archetype associated with the number 12.

Count its expressions: Pure light is made up of 12 colors. Each human cell has 12 biochemical salts. Photosynthesis is arranged in a complex twelvefold symmetrical pattern. The number 12 recurs in the Bible* and is also associated with Buddhism (the 12 Nidanas to create samsara), Judaic (the 12 tribes of Israel (Genesis 49)) and Islamic religions (the Twelve Tribes of Israel (Surah Al-Baqarah—60 and Surah Al-A'raf—160 in the Quran), 12 astrological signs, the 12 Chinese zodiacs, Greek and Egyptian mythologies (The Labors of Heracles was increased from 10 to make 12, and Sanskrit[6] legends. Many things in nature arise with a twelvefold symmetry. Twelve is the number of completion and wholeness. Because of the vertex arrangement of the fourth Platonic solid, the dodecahedron, the rest of the Platonic solids—tetrahedron, cube, octahedron, and the icosahedron—can be inscribed within the dodecahedron 12 pentagons, 20 points, and 30 edges. This arrangement reflects the dodecahedron's total and holistic nature. (See plate 4.)

The earliest example of a dodecahedral instrument dates from the Roman Empire. Made of bronze, it was about four inches in diameter

---

*Jesus Christ chose 12 disciples and replaced Judas Iscariot with Mattias to maintain 12 disciples (Acts 1); 12 unleavened cakes were placed in the Tabernacle (Leviticus 24:5); 12 silver plates, 12 bowls, 12 bulls, 12 rams, and 12 male lambs were instructed to be placed in the Tabernacle (Numbers 7), and so on.

and speculated to have been used as a measuring device, military range-finder, or even a fortune-telling tool based on comparable examples found in Gallo-Roman sites.[7]

The dodecahedron has found its way into nearly every aspect of our universe, whether looking in or looking out. For example, only a handful of planetariums existed before 1940 because they were very expensive to make. Armand Spitz, charged with building an optical-mechanical class* planetarium based in Chadds Ford, Pennsylvania, struggled to assemble the dome using an icosahedron shape.[8] Only after Albert Einstein suggested using a dodecahedron for the planetarium star projector did Spitz find success: his model became popular and over a million units called *A Spitz Junior* home planetarium projectors were manufactured from the mid-fifties to the seventies, showing mechanized motion for the sun, moon and planets, and lunar phases.[9] Similar to Spitz using a dodecahedron model to observe celestial objects, DNA's dodecahedron shape further reveals the relationship between genetic strands and the Universe.

Think of the dodecahedron as a key. This key unlocks memory in terms of behaviors, mutations, gene frequencies, gene selection, and inheritance. It incorporates neo-Darwinian and Lamarckian inheritance, trans-dimensional physics and biology, Jung's archetypal approach to the collective unconscious, and both Pythagorean and Platonic logic. To absorb such a key, we have to free our mind from the original double helix.

One of the first people to reveal hints of DNA's dodecahedral form was the researcher Robert Langridge. Considered a pioneer of molecular graphics, it was 1957 when he earned his Ph.D. in crystallography, with a dissertation on X-ray crystallography, model building, and computational studies of the structure of DNA. Langridge studied and was guided by Maurice Wilkins, his doctoral advisor. The diagram of the DNA structure that is pictured in Watson and Crick's article, "Molecular Structure of Nucleic Acids," was drawn by Langridge. "The picture of DNA that's in that article, I drew with a pen."[11] Over the next several decades Langridge continued his structural investigation of

---

*The optical-mechanical class initiated as the A-series. The A-series developed into the 512 series with the advantage of using digital voltages to the projector.[10]

DNA, eventually producing the image of a cross section of DNA that reveals a ten-sided geometry. (See plate 5.)

The dodecahedron is a harbinger of new science. Consider proteins, whose individual components, called monomers, can be linked together to form a chain, or polymer. A 12-monomer protein naturally arranges itself into a dodecahedral shape. Its morphic field has the behavioral memory to model itself according to its memory of how DNA is arranged. Yet, this memory of the dodecahedral shape looks very different from what we think of as the shape of the traditional double helix. That DNA can take on various configurations with nontraditional base pairing is a matter of no small interest. But beyond the way different base sequences are grouped, mismatched, or assimilated, DNA's implicit structure represents *remembrance* and *storage*. In this context, I mean memory as a construct of origin and essence, much like the divine source that inspires our thoughts. We are now just exploring quadruple helix structures in healthy living cells, and it is this level of shift in perspective I want to initiate inside of us, using our own thought processes. (See plate 6.)

The dodecahedron shares a fivefold symmetry and reciprocal geometry with the icosahedron, the Platonic solid representing the form of water. Both forms have been recognized in the modern fields of microbiology, astronomy, and crystallography. The structural understanding of these forms is not only important but necessary, as they serve as blueprints to emerging health and scientific interventions for the material world. For instance, the head of a bacteriophage and a considerable number of viruses—pox, measles, HIV, herpes simplex virus, influenza, norovirus, adenovirus, hepatitis B, to name just a few—share an icosahedral architectural capsid. The ancient Greeks used the dodecahedron overlain on the icosahedron to create a grid of 120 triangles. When laid over a globe, they mapped and measured their edges and vertices to predict astronomical events such as equinoxes and solstices where the triangles coincided with mountain ranges and boundaries of tectonic plates.[12] Similarly, with DNA as a complex geometric projection of a dodecahedron, the icosahedron naturally instills the braiding and spiraling atomic movements to incorporate DNA's self-assembly, helical stabilization, design, and morphology. To confirm this theory in practice, we already know a well-defined spine of hydration, a zigzag string of water molecules, make up the integrity of a double helix.

Because the icosahedron and dodecahedron share a reciprocal edge symmetry yet opposing face and vertex symmetries, this complementary and reciprocal geometric relationship enables them to transcend energetic properties. The basic molecular geometry of water is tetrahedral. Due to its intrinsic fourfold rotational field, the icosahedral network is composed of 14 tetrahedrons arranged into 20 clusters and 280 water molecules (20 × 14 tetrahedral units). We know this because an icosahedron has 20 faces. The dodecahedron contains 20 vertices whereby atomic spinning occurs at the vertices. On an atomic level, the edges of the internal tetrahedrons spin to form a dodecahedron and fluidly change positions within the icosahedron based upon the changes in hydrogen bonding, electrostatic interactions, and hydration of the cluster. While the relevance of spin symmetry will be further explained in chapter 7, the most critical highlight in this section is the hexagon-pentagon connection, the ability for the dual pair to shift from one shape to the other. The icosahedron-dodecahedron dual pair can be seen abundantly in the chemical world. In chapter 7 we will explore in detail how water interacts with DNA.

My First Observation of a Dodecamer

During embryogenesis, Danio rerio, *commonly known as zebrafish, has a similar cell-signaling mechanism as humans, making the species an ideal model to study human genetics and development. As an undergraduate research assistant in Dr. Robert Tombes's embryology laboratory, I examined the cell morphology and motility of early zebrafish development, particularly the morphological changes in each cell during embryogenesis using confocal microscopy.* With it, one can see a cell's shape and, depending on the experiment, its activities.*

When a protein folds, its molecule assumes a unique structure, essential for its biological function and memory of the protein-DNA

---

*Confocal imaging uses a laser and two pinholes. The pinholes work to exclude any photons that may be coming from a different focal plane and result in capturing clear images of extremely thin tissue sections, less than 1 μm.

complex. A protein kinase called CaMK-II (Calcium/calmodulin-dependent protein kinase II), an enzyme that plays a key role in calcium signaling in eurkaryotic cells, takes on a "supra-molecular" organization.[13] *CaMK-II is made up of 12 subunits that are assembled into 2 rings of 6 subunits each. Each subunit is encoded by a distinct gene, which can yield different isoforms or protein variants through alternative splicing. When I looked at the structure through the confocal, I saw a dodecamerous hub composed of two hexamer rings stacked one on top of the other. It was the first time that I saw the dodecamer in a cellular environment, let alone in nature at large, and I intuited how deep and significant it was, though it would take me years to process the intuition and to realize that it had possibilities for me to tap my own deeper energy skills.*

DODECAHEDRAL PLATONIC FORMS INFORM CONSCIOUSNESS

The dodecahedron summarizes Platonic characteristics of Forms; here, I outline its components in relation to consciousness:

The dodecahedron symbolizes the Universe as a Platonic solid; its archetypes unite all human experiences. It is associated with aether, consciousness, and quintessence.

The dodecahedron has perpetual patterns; the phi ratios in its pentagons are seen all through nature from plants to insects to animals and the human body.

The dodecahedron is all-encompassing; it is able to contain all of the other Platonic solids within its Form. It shows talent in exhibiting different tendencies, characteristics of nature, and interconnectedness.

The dodecahedron demonstrates aspects of nondualism and dualism by its coupling with the icosahedron. The relationship between dualism and nondualism can be approached through two closely connected concepts: consciousness and awareness. Although related in context, the differences between consciousness and awareness can be distinctly considered. Consciousness can be a dualistic and embedded cognitive process, whereas awareness is a nondual and nonlocal process.[14] In the field of neuroscience, cognitive consciousness is processed by the neocortical and pallial circuits, whereas the nonspecific, nonlocal awareness is processed by the precortical (subcortical) circuits. The dodecahedron demonstrates both dualistic and non-dualistic qualities; its parity with the icosahedron shows its dualistic properties as a representation of consciousness. Conversely, its nondual feature, or simply being an independent Platonic solid, postulates its unifying nature and emphasis on being grounded as a singular principle among a heterogeneous pool.

The polytope of the dodecahedron is timeless and spaceless; hence, its fourth dimensional properties access a realm unconstrained by space and time. The intelligent realm that supports the Universe and the existence of all other elements is the source of consciousness from a higher dimension. Consciousness created matter, and it came first.

The principles of Universal consciousness are unity, infinity, and harmony, which are evident qualities of archetypes, thus, further exemplifying that humans carry archetypal DNA.

BENDING, EVOLUTION, AND RETRACING THE PAST INTO TOMORROW

A line extends infinitely in all directions, but when it intersects another line, they become interlinked. Space-time has no basis for straight lines. If light bends around galaxies, information bends in DNA. Water (icosahedron) and DNA form a dual pair, interweaving with one another spontaneously. The polar qualities of a water molecule and DNA's electromagnetic field causes the double helix to bend beyond the configuration of a spiral.

Plato wrote in *Timaeus*:

In almost all we have said we have been describing the products of intelligence; but beside reason we must also set the results of necessity. For this world came into being from a mixture and combination of necessity and intelligence. Intelligence controlled necessity by persuading it for the most part to bring about the best result, and it was by this subordination of necessity to reasonable persuasion that the universe was originally constituted as it is. So that to give a true account of how it came to be on these principles, one must bring in the indeterminate cause so far as its nature permits. We must therefore retrace our steps, and find another suitable principle for this part of our story, and begin again from the beginning as we did before.[15]

The time has come to rethink the intelligence of being, space, and becoming. We are, in a sense and according to Plato, *supposed* to begin again from the beginning after allowing the pieces to fall into place naturally, which is distinguished by the act of retracing. *This* is evolution.

THE MISSING HERO
A Jungian Archetype

In the formal expression of recognizing the Jungian tradition in analytical psychology of archetypes, I present an archetype that resonates with the 12-stranded DNA: the missing hero. Joseph Campbell was the first to articulate the common motif of the hero's journey, influenced by Jung's archetypal principles. It is a masculine interpretation of a universal person setting out on a mysterious adventure and triumphing, sometimes miraculously, in each crisis. When the person returns home, his mind and being are transformed. However, there seems to be a lesion and gap: it is the missing hero.

The missing hero is the bridge between the masculine and feminine, the connective network that unites all archetypes and templates, because the missing hero both is and isn't a character. It is a timeless, felt experience. The missing hero uses imagery, both light and dark, that is beyond duality or good and bad.

My heart aches for the missing hero, always aglitter with curiosity and fascination, always cheerleading the collective, always showing multiplicity. When we see a divine guide in our waking or sleeping vision, the personified inspiration captivates our hearts and minds. The character becomes the main theme, protagonist, or hero of the dream or story. The hero has come to save the day; he or she is courageous, willing to protect others and self-sacrifice to stop the world from ending. The hero takes over the story and distracts from other characters who also share a vital role in the story's success. That's a weakness of humanity that we see in communities and cults—too many heroes and not enough lovers. The good news is that human nature is designed to cultivate inclusivity, so we can, too.

Perhaps the protagonist we seek is more subtle than the hero. Perhaps a combination of necessity and intelligence shifts our attention from the hero himself to include the missing hero: the various heroic antiheroes and all of the other players on the team. By listening to our own thoughts, ideas, and dreams, we can paradoxically steer away from a self-centered ego, saving itself and the world for a vaster story that centers around harmony, compassion, collaboration, and creativity.

This book spotlights the missing hero, the antihero, the shadow hero, the orphaned or exiled hero; it is its own innocence project in the sense that it restores degrees of freedom that we have lost under false crimes and acclimations to the unfolding conditions of the emerging world. The evolution of our DNA likewise isn't linear or split into algorithms, methods, or codes. The antihero guides each next step forward in our evolution. Our 12-stranded DNA is our missing hero, helping us understand that our ethereal evolution is simultaneously spiritual, physical, and elemental. We have both the practical and spiritual skills needed to touch the unknown through courage and imagination.

As each archetype offers its innate wisdom, it's important to acknowledge our gratitude. We express our thankfulness to the circle of life, recognizing that each dream figure, image, or character in our waking and slumbering experiences is an evolutionary catalyst, a hero or antihero.

As we progress, we naturally cheer others on—other humans, creatures, spirits, and tribes—we don't stay concerned with winning or status. This tournament has always been collective: our higher selves, better spirits, and shadows against the darkness of the uncreated, the hell realms of self-sabotage, self-bondage, and anger-fear. Our clairsentient lens focuses on seeing that love can't be confined to a single character, definition, or accomplishment. We embrace our own and others' distinct qualities and see them as the reason why life is diverse and homing—a niche or biome—for all.

The missing hero wants to share, rejoice, and fission-fuse in everyone's current state. The missing hero celebrates creation's diversity. It claps for everyone, no matter who or what. It is even as William Blake asked of his Tyger: "Did he who made the Lamb make thee?"[16] Our greatest gift is being awake to this love and giving it back to the world.

We are the missing hero. We are able to experience, assimilate, and transform anything.

∽∘∽

I left corporate science for many reasons, one of which was to write this book and also to acknowledge a deeper sense of who I am, which resonates more like the mythological legend of Longmen, or "The Dragon Gate." In Chinese tradition, this tale is a story about carp who were brave enough to leap over the cascading waterfall from a mountain and transform into a dragon.

I am yet more intrigued by the forceful water that brought many carp down the river, unable to swim back—leaving them fixed, static, and stuck in the form that they are. This aspect of the story reveals the plot's confronting dynamics, the movement and resistance of evolution simultaneously. Lingering since the origin of form, this very topic is an obscure, yet critical theme in each person's life, on an individual and cosmic basis.

My hero's journey began when I was born, to Michail Remennikov and Stanislava Remennikova in Kyiv, Ukraine in 1988. At the time, I was too young to remember Ukraine's path to independence, but its restructuring or perestroika is the reason why my father decided to immigrate to the states in 1990. In addition to avoiding the Chernobyl radioactive catastrophe and mounting economic problems, he was also confronting a bigger problem—an overtly political agenda involving the autonomous self-organization of a nation and the bitter resistance toward it from the communist members of the former Soviet Union.

As a young child, my father's decision to move, a prescient one for that matter, instilled in me an ideological and practical grasp of autonomy, authority, and intention. I leaned into my education and career track driven by the notion that it would redeem my family's pursuit for financial stability and be a marker of success to my parents.

Following the devastating and unforeseen loss of my father, I embarked on a drastically transformative year—physically, psychologically, spiritually, sexually, and nutritionally. As I trained for an Ironman, navigated a divorce, and left a seven-year religious cult, it felt as though I was undergoing my own perestroika and path to independence, mirroring the revival of Ukraine and the fresh start my father had been seeking. Perhaps, the missing hero is the action of trust, hope, and courage. What, then, is the relevance of this forceful water to me, and did it shape me to become the protagonist or antagonist in my own story?

Revealing Our 12-Stranded DNA by Pythagorean Gematria

Numbers are the most distinct liaisons between humanity and creation. Numbers differentiate, quantify, and solve problems. A number is a precise, accurate and honest representation of vibration, hence the interfacing model of existence. Gematria is assigning numbers to words or phrases according to an alphanumerical system, through which the meaning can be decrypted, revealing thought-like information from the cosmos. Decryption converts ciphertext to plaintext: number to message. Encryption is the reverse, the enfolding of original information to an unintelligible code. Both are necessary for genesis, creation.

My practice of gematria is based on the alpha-numerology of the Hellenic language, by which I decrypt unseen knowledge from vibrations or forms using number intelligence. My system combines sound, math, frequency, vibration, and symmetry with traditional science so that intelligence can transcend and ascend.

My discovery of gematria began with my spiritual network. In 2016, I started studying astrology and took classes from professional Nick Lasky. In conversations about numerology, I shared my desire for additional knowledge and training; I was already reading numbers

intuitively and wanted to learn more about number theory. Nick and my psychic friend Julia Ennis introduced me to Mega Pythagorean and Greek mathematician Eleftherios Argyropoulous to get my name "read" or decoded using the Pythagorean alphanumeric system of the Hellenic language. At the time, I didn't understand what they meant but decided to follow along. After a two week's wait and submitting my first, middle, and last name along with my date of birth and location and my parents' names, I received pages and pages of information of who I am incarnating, my purpose and missions, past lives and destiny. Before I go any further, you have to understand: I am a trained analytical research professional by trade. There's an inherent part of me that has always been a skeptic. However, when I read through the scrolls and the story of me, I instantly understood that I am supposed to learn and integrate this method.

This became my new science, formula building, and a significant aspect of my thought process. The one-on-one classes that I did with Eleftherios were mystical, revelatory, and introduced me to a new form of theosophy. The numbers shifted my lens on the world. Numbers reveal an imperceptible and invisible wisdom.

I want to make an important distinction here. Biotechnology uses traditional science, while I use gematria as a traditionary science. Traditionary sciences include astrology, magic, numerology, and energy medicines like shamanic transference, homeopathy, and cranial osteopathy. I believe in them because they are adapted from the most primitive form of sound that our earliest ancestors assembled, organized, and determined in the first phonetic language (before that, they likely had whistles, drums, and throat songs). Homeopathy, cranial osteopathy, and sound healing are vibrational medicines that address the electromagnetic signature of molecules, cells, tissues, and organs. The rationale for each alternative medicine also highlights a unique approach toward exploring sensitivity to electromagnetic fields. The basis of homeopathy incorporates a water system by using a series of dilutions to activate the potency of a substance. Cranial osteopathy utilizes a bodywork therapy that augments vital bio-communication in a patient's fascial system. Similarly, sound healing is an energetic phenomenon that applies particular frequencies to the body. In each modality, the common theme is to restore an imbalance within the system by introducing a signal that cancels a pathological frequency that is causing a deficiency, alteration, or excess molecular interaction in the body. As a similar principle in pharmacology, this intervention toward normal functioning is approached though a drug. Gene therapies and energy medicines are all built upon the same evolved language, which leads me to a necessary, universal proposition that is appropriately composed by Cyril Smith: What is urgently needed is to be able to read the language of electromagnetic bio-communication to complement our understanding of the genetic code.[1]

To speak to its significance and our understanding of the fundamental nature of all matter, let's explore gematria.

REVEALING SPIRITUAL MESSAGES USING GEMATRIA

Before describing my system, it is worthwhile to see how these ancient codes are still speaking. Jay Kennedy, a science historian from the University of Manchester, discovered a code in Platonic texts and took a shot at their secret messages. These, if authentic, were well hidden by a musical code concealing Plato's symbolic system.

In the journal *Apeiron*, Jay Kennedy proposed that Plato used an "order form" of the symbols that he acquired from the followers of Pythagoras, giving his books a musical structure. He discovered Plato's placement of words related to music were after each twelfth of the text in *Republic*. This structure depicts the dodecaphonic scale of twelve notes of the ancient Greek musical scale. Certain notes were harmonious and pertained to love or laughter, while others were inharmonious and related to war or death.

Plato's camouflaged system may have been based upon the so-called "harmony of spheres" that Pythagoras declared a century earlier. According to Pythagoras, the harmony of spheres refers to the non-acoustic music created by the movement of the planets and constellations. "The books of Plato played an important role for the foundation of Western culture, but they are mysterious, and they end with puzzles," Dr. Kennedy explained to the faculty of Manchester in Biosciences.[2] He continued:

> I have shown rigorously that the books do contain codes and symbols and that unraveling them reveals the hidden philosophy of Plato. This is a true discovery, not simply reinterpretation. This will transform the early history of Western thought, and especially the histories of ancient science, mathematics, music, and philosophy.

As a reiteration of the salient piece from the University of Manchester's press release, I am getting down to an interesting, fundamental concept. First, the hidden codes, if more than ex post facto gaming, indicate that Plato prefigured the scientific revolution nearly two millennia before Isaac Newton's era. Notably, Plato established that "the book of nature" is not only written in the language of mathematics, but is also a means to approach the creationary, interpretive methods of the gods or God. Science and religion are the same algebra, geometry, and gematria. However, Plato did not encrypt texts for pleasure or recreation. For a meta-Pythagorean, engaging in sacred secrecy might result in a death sentence. His knowledge and teachings that the universe is regulated by mathematical laws and not by gods on Olympus were a threat to ancient Greek theology. Socrates, Plato's teacher, had been executed for less: for organizing a sect.

GREEK LANGUAGE: THE HELLENIC LEGACY
A Brief History

The history of the Greek language must be considered before I discuss Pythagorean gematria. The ancient Greek and Chinese languages are still known to us after 3,500 years, along with other languages of antiquity—Sumerian, Egyptian, Hebrew, or Arabic—but Greek can be regarded as the most influential language in the world.[3] Even when the Romans began to conquer the East beginning in the first century BCE, the Greek language continued to be spoken, and it influenced the Latin language from the second century BCE. With just over 7,000 spoken languages in the world, Greek letters are universally used in mathematical and scientific notations, such as physics, trigonometry, and engineering. Although history of the Greek language is a separate topic, I will highlight main points that transcend linguistics and bring us to gematria.

The "Greek" language is a descendant of the final phase of Indo-European expansion in Europe. In the complex of Indo-European languages, the Greek language developed from the polythematic Indo-European, or the Proto-Greek language, the starting point of the history of the fragmentation of Greek dialects. Traditionally, the Minoan civilization of Crete predate the Proto-Greek speakers in the Southern Balkan peninsula (now the territory of the Hellenic Republic) by over a millennium. When the first speakers of Proto-Greek inhabited this territory, language was the most significant and fueling medium that aggregated and divided the people into tribes. The tribes gave way to the Mycenean civilization, or the memory associated to Classical Greece (approximately 600 BCE and 400 BCE). During this stage, classical Greek scholars led influential and innovative works in the fields of mathematics, philosophy medicine, astronomy, literature, and the development of language and alphabet.

Modern day Greece is also referred to as the Hellenic Republic; the translation of Greece is Ελλαδα, Hellada or Hellas. As a result of Roman influence, the term Greek comes from the Latin *graeci,* which became the common descriptor of Greek culture and people. In the context of this book, the word "Greek" is intentionally treated as a legacy and part of the Hellenic language. One of the aims of this book is not to particularly elaborate on the distinction between *Greek* and *Hellenic,* but to emphasize the compelling gift that Hellenism has modeled for the modern world. That is, a deep calling in remembering our identity. Not only to the inexhaustible question, *who am I?* The gift of Hellenism is a recollection of our heritage, which continues to transform in and of itself. It is the way we are reminded of being open-minded, accepting truth and beauty, and evolving ourselves. Even non-Hellenic people have integrated Hellenic values and ideas. The seminal legacy of Hellenism is the heart of what our DNA will mean to us going forward.

Historically and culturally, the Greek language has often been interchangeably translated into other languages. Its linguistic features were adapted to write languages such as Latin, English, and the Slavic languages, and most importantly, its key aspect is evident in its evolving dialects, from Mycenaean to Hellenistic to modern Greek today. Its alphabet, phonology, morphology, syntax, and lexicon have hidden hermetic codes and tendencies. The ancient Greek alphabet was adapted from the first fully phonemic script, the Semitic alphabet of the Phoenicians.[4]

Morphophonemics generated a system of compounding, inflection, phraseology, and phonetics, among other multifaceted linguistic innovations. Inflection is modifying a word to place it in the grammatical structure of a sentence. The Greek vocabulary not only proliferated new abstracts and terms for word derivation or verbal compositions, but also introduced a *means* to approach derivation and compositions using its well-developed morphologically rich system. Along with its syntactical borrowings, this concept can be seen in scientific writing (e.g., amphi, archaea, arthro, bios, etc.).

Before the development of the Hellenistic era, Greek genre implied dialects. Depending on the provenance or type of literature, the Attic, Ionic, Aeolic, Doric, and Koine were among a number of dialects that contributed to the spiraling evolution of the ancient Greek language and culture to the dialects and language that are spoken in the modern era. Hellenisms are not only foreign words which were later absorbed and utilized in different languages, but also were a fertile source of formative elements such as roots, suffixes, prefixes, and methods of word composition. As the Greek language survived between time periods, a number of linguistic structures varied. Among the ones pertinent to the book is the gradual loss of long and short vowels and the shift in ancient pronunciations of certain consonants (β, γ, δ, φ, θ, χ and ζ. Β, Γ, Δ).[5] In this sense, the underlying morphological archaisms are a dynamic, integral component of the phrases that I reveal in the next section.

To explore the 12-stranded DNA, there is a proliferation of new compound words, new phraseology, and a daring syntax that I incorporate using an independent tradition. I treated my inquiry of these phrases not from an angle of technicality or scientific dogma, but as a well-mannered observer and courteous student of the arts, sciences, and culture of Hellenism. The Hellenic language (a term used to describe the implicit group of languages that descended from the ancient Greek language), as an original code, is effective in repurposing stem words to disclose a variety of new meanings. By using combinations of basic words and interchanging signifying syllables, we birth many new ideas and inventions we might not have hit upon otherwise. Divination is literally divine transmission.

To examine the unique intelligence of gematria in regards to the 12-stranded DNA, refer to "Gematria Meditations" in chapter 16. Chapter 17 explores a set of sound frequencies that pertain to the dodecahedron.

Gematria is a system of assigning numerical values to letters, words, or phrases bearing the notion that words or phrases with identical mathematical values could indicate a relationship with one another in the same way symmetry reveals a proportionate similarity between two halves of an object. Think of it as a mirror. When converted to a number, the identical mathematical values associated to the words and phrases may be compared to gain insight with interrelated abstractions and finding correspondences between words and concepts. Although the origins of this system is ostensibly derived from the ancient Greeks (gematria is the Greek term *geometria*, "measurement of the Earth" or "geometry"), gematria is an application widely used in Hebrew texts (notably associated to the Kabbalah). Analogs of this system are seen in Arabic and English languages.

Molecular Gematria

DNA Strands and Morphic Evolution

Now that we've seen the potential and mysterious insight of gematria, we can apply these ideas in a special way to reveal how DNA propels humanity's evolution. We will see how the structure of DNA creates morphogenetic fields (or "morphic fields"), acting nonlocally with breathtaking storage capacity while connecting us to the cosmos.

DNA STRANDS

We saw in chapter 1 how the traditional DNA helix comprises two linked nucleotide chains, or strands. To explore the esoteric nature and capabilities of the 12-stranded DNA, we will begin by reevaluating the term *strand* through the lens of gematria. If you think about the word "strand" as a palindrome and read it backwards, you have the letters DNARTS. Splitting the word in the middle, I deciphered "DNA" and "RTS." We know DNA is the description of the genetic material, deoxyribonucleic acid. I designated the three letters on the left—RTS—to represent a new concept of the function of DNA: Rapid Technology Storage.

STRAND = DNARTS = DNA + RTS

A coincidence? I digress to examine the importance of the word "strand," which is central to my book and the relevance of this coincidence.

Examining the word through the palindromic lens was an essential stepping stone to import my message. It uncovered the primary focus of this book, DNA. That became the fundamental coincidence in this current analysis. Let's define coincidence. A meaningful coincidence is the coming together of two incidents in a surprising and unexpected way that has meaning to the person observing it. This first meaning is very striking, both the book and palindrome have the same letters: DNA. Amazing!

Besides Carl Jung, one of the giants who has expanded our understanding and depth to the term "meaningful coincidences" is through the lifelong works of Dr. Bernard Beitman. In his book, *Meaningful Coincidences: How and Why Synchronicity and Serendipity Happen*, Dr. Beitman characterizes meaningful coincidences by three forms of meaning: the emotional response of the observer, the applications to the observer's life narrative, and clues to how the coincidence might be explored. I was very surprised, intrigued, and curious about the DNA semantic parallel. I decided to keep going. What might RTS mean? In the context of the dodecahedron as the two-stranded DNA model, RTS easily becomes an acronym for a vital DNA process, rapid technological transfer and storage, which is an apt description of how DNA operates. Finally, how might this coincidence be explained? While there are numerous explanations being offered to answer questions like this, I suggest that my attempt to uncover the underscoring truth about DNA is being reinforced by this coincidence. Dr. Beitman emphasizes that when we are in flow of pursuing answers to major life questions, we are also in the flow of experience a correlating increase in meaningful coincidences.

So what is "Rapid Technology Storage?" In the following sections, we will look at the meanings I intuit from the gematria of RTS.

RAPID

A Self-Organizing Morphogenetic Field and Instinctive Memory

DNA has a bidirectional orientation, consisting of magnetic and electric fields that spiral around one another, creating a geometric wave pattern that repeats itself within itself, a fractal. Fractals exhibit similar patterns at various scales demonstrating a property known as self-similarity either by size or time scales. These complex structures are geometric shapes that can be split into parts, each of which is a reduced-size copy of the whole. The key element about fractals is they are generated by recursive processes, which ties back into recursion itself. This process not only repeats ad infinitum, an endless cycle of self-duplication yielding layers of complexity that emphasize the interplay between chaos and order, but also highlights an esoteric emergence of continual "self" consciousness through an archetypal model.

Before we explore the role of the archetype, its origin must be considered. The term *archetype* comes from the Greek word *arkhetypon*, and its original meaning, "beginning pattern," "model," or "figure on a seal," should be distinguished from its Jungian version of "image or idea from the collective unconscious." In fact, the neuter of the adjective *arkhetypos* means "first-molded." The Greek word for rapid is ταχύς (*tachýs*), which represents a gematrical sum of 1501. It shares the same alphanumerical sum with the word ἄρχω (árcho), which means "to be the first."

The meaning of DNA can be felt as a cosmic intersection in which archetypes are constantly recreating themselves based on the influence of their individual past states.

An archetype is distinguishable depending on the resolution of the fractal. If the resolution is high, the archetype is endemic to a situation or event. If the resolution is low, the archetype is more abstract and may diffuse relatively over time. The number of archetypes a person comes across within themselves and in nature is unlimited.

Fractals are also self-organizing morphic fields containing evolutionary information; these memories are a fundamental property of the genome's nonlocality. Biological "memory" can be seen when a body regenerates a lost part: planaria and salamanders are able to regenerate some of their tissues and organs after amputation, and there is also wound healing in mammals, regeneration of antlers, liver, and fingertips.[1] Morphogenetic fields, proposed by Rupert Sheldrake and pioneered by Alexander Gurwitsch, describe complex, interconnected energy systems guiding cellular organization, development, and coherent responses to biochemical signals. Morphogenetic fields are as real as gravitational, electromagnetic, and quantum matter fields. The existence of a morpho-genetic field or a "morphic field" is an intrinsic property within every cell, tissue, organ, and living organism. A morphic field contains the memory for the system it organizes and suggests that evolutionary information appears to be an emergent property of larger fields that permeate both time and space, as Einstein suggested. It invites the perspective of the "laws of the universe" to evolve over time, instead of regarding the cosmos has an eternal machine—a human-made concept.

Within our DNA, fractals weave layers of morphic fields that connect our genes with celestial patterns in the cosmos. This field is transmitted from past members of human and mammal lineages through morphic resonance. In a sense, morphic resonance conveys instinctive memory.

An instinct is an intricate electrochemical impulse. Instincts are complex because they are not present at birth, they are attributable to genes, not learned, and they are adapted during evolution.[2] While there's a tremendous chasm between assumptions surrounding instinct and its technical scientific explanation, my aim is consistent with its morphic resonance originating from our archetypal DNA: I believe instinct is embedded in our *noncoding* DNA.

Noncoding DNA

While the human genome contains 3.3 billion nucleotides, 3.2 billion of them do not get translated into proteins, and these are called "noncoding" DNA. Or, what science historically referred to as "junk DNA." The term was first used in the 1960s and was formalized by biologist Susumu Ohno in the 1970s. Ohno argued the material inevitably accumulated inactive sequences over many generations. In a *Nature* review published in 1980, Francis Crick stated that "junk DNA" had little specificity and conveys little or no selective advantage to the organism.[3]

Its initial "junk" descriptor discouraged people from researching it any further. As we move toward an Aquarian epoch, scientists intuited the hidden treasures within this previously discarded system, and the term "junk DNA" slowly faded and transitioned to "noncoding DNA." Noncoding DNA doesn't code for protein, but it has an equally important role.

Noncoding DNA contains segments that can change their location within a genome; these are called transposons, or transposable elements. They've acquired the name "jumping genes" and can make many copies of themselves throughout the genome. Their transposable elements are a function of gene memory. About half of the human genome is made of transposons,[4] influencing the genome's evolutionary trajectory through genetic and epigenetic variations. This rich content in "junk DNA" is like gematria and *isopsephia* at a molecular level in the way it contains mystifying and meaningful information that relates to our evolution. *Isopsephia* is a term that indicates equivalent gematria value between two separate phrases or words. This form of parity phenomenon may appear to be coincidental, but can also be a tool in bridging concepts with a higher consciousness. The application of isopsephia is explored in chapters 16 and 17 where it can be an access point to tap into revelations and an intersection for meaningful ideas. For example, the Greek term for synchronicity, or συγχρονικοτητα, shares an isopsephia as the the cumulative gematria value of αηρ υδωρ πυρ γη αιθηρ, or air, water, fire, earth, aether, which is 2132. The probability, given the relevance of their respective meanings, between these two structures (one being a single term and the second application treated as one collective term or group) having the same isopsephia is low. Gematria and isopsephia can be an enlightening way to observe reality through a numerical standpoint.

Defective Genes Can Evolve into Something New

Pseudogenes are non-functional remnants of once-active genes. However, as DNA intelligence reemerges with molecular memory, pseudogenes evolve with new functions and find second lives. If their RNA is similar enough to that of a working gene, they can control protein activity. A study by researcher Manyuan Long using a major food-crop species of rice—the *Oryza* species—showed that random, noncoding DNA can even evolve to produce new proteins.[5] According to Dr. Long, "Using a big genome comparison, we show that noncoding sequences can evolve completely novel proteins. That's a huge discovery."

For a long time, scientists believed there were two major ways new genes evolved: duplication and recombination, but now they wonder about a third way where genes evolve from scratch. Is it possible that the noncoding sections—an astonishing 97 percent of the total genome—can acquire mutations or variations that make them functional? Simply put, evolution is not possible without genetic change, and mutations are essential for evolution.* While understanding the noncoding part of the genome would enable further understanding of identifying genetic mechanisms of disease, it could also introduce a deeper learning method into actionable results for personalized holistic wellness. Emerging evidence suggests that adaptive evolution is not restricted to protein-coding DNA. Particularly, the adaptive evolution in human noncoding DNA seems to be more common in the regulatory region of genes involved in cognition and metabolism compared to genes in other processes.[6]

In the next couple of sections, I will dissect how a morphic field folds into an aspect of DNA's capability of being regarded as a technology.

When thinking about my journey, it's interesting to me to think back to the days when I was excited about my role in traditional science. It all started about six months after graduating, when I found the job that led to a decade-long career at a pharmaceutical and biotechnology contract research organization, Pharmaceutical Product Development. What began as an entry-level bench chemist position developed into my becoming principal investigator and subject matter expert on a number of different projects, including norovirus, dengue virus, and coronavirus disease. There were various departments and new protocols that I was curious about and committed to learn more about, including regulatory affairs, drug development, data management, biostatistics, scientific writing, compliance, and project management. I had an insatiable thirst for knowledge, experience, and growth.

When I first became a part of the company, the basic immunoassay technique that I trained on was an enzyme linked immunosorbent assay, or ELISA. I learned and practiced this method of measuring

---

*Genetic mutations can have varying individual effects; some mutations are proven to be an advantage for the organism's survival and adaptability to their environment. The most significant and relevant mutation in the context of this book begins with the change of a person's mind and thoughtform process, which is ideally prompted by desire and recognition of evolving oneself with deliberate consciousness.

and detecting proteins, antibodies, and antigens, examining the immunological response against bacterial and viral infections.

In my fourth year, I transitioned to the vaccine sciences department to work on a larger clinical program for norovirus and a separate pneumococcal project. During this segment of my career, I was promoted to senior scientist and dedicated more time to research and development. A typical workday for me during these years would be to validate and automate a bench assay with a 50 percent pass rate and improve it to a 90 percent pass rate, which I successfully did.

TECHNOLOGY
Quantum, Nonlocal Attributes of DNA

Approximately 13 percent of the human genome can fold into noncanonical DNA structures. Plate 7 shows the differences in these conformations, with the traditional Watson-Crick double helix, known as B-DNA, appearing in the middle.

The relationship between all DNA variations, loci, mutations, and substitutions can be captured through allele frequencies. An allelle is a specific version of a nucleotide sequence at a particular locus or location on a gene. For instance, scientists can examine allele frequencies in a nucleotide substitution to map how a genome spans different evolutionary times, which enables us to savvy how genes produce the vast diversity of modern-day life forms on our planet. Although DNA's conformation (the most common form of DNA known is the canonical right-handed or B-DNA) depends on its underlying sequence and stability, its conditions within the cell are extremely significant for its morphology. Supercoiling, transcriptional activity, and ionic concentrations in the nucleus are all interactions that can influence the conformational transitions and genetic variations between B and non-B DNA.

It is challenging to study the transient nature between DNA loci that form non-B DNA structures at a given time. I believe the observation of a biophysical stimulus such as proliferation, migration, cellular differentiation, and synthesis of proteins is influenced by the cell membrane's morphic field. Its morphic field can be fundamentally generalized in this way: A cell membrane is organized by lipids (phospholipids and cholesterol), proteins, and carbohydrate groups. Each cell in our body is encased in this tiny bubble of membrane, which serves to take messages from its environment and convey it to the internal structures of the cell such as the nucleus (DNA), manifesting algorithms that organisms use to do their respective jobs of repair and damage control. The cell membrane's importance cannot be underestimated: without a membrane, you don't have a structure at all, just a bunch of organelles in a mineral pool, let alone complex multicellular life forms!

Chris Jeynes and Michael Parker have mathematically described this morphic field of information and identified its entropic influence by visualizing geometry forms in nature, such as the galaxies' strange symmetry and the symmetry found in double helical spirals. Entropy is the measure of disorder of a system, and, entropic influence is characterized by disorder *always* increasing in nature. They described this new concept that links information and entropy together as holomorphic info-entropy—much like conglomerating space and time as one manifold of space-time or electricity and magnetism into electromagnetism. Jeynes and Parker have also described how a change in state will result in a change in another state. Therefore, a change in information or entropy results in the same reciprocal influence through geometry:

> We showed that entropy and information can be treated as a field as they are related to geometry. Think of the two strands of the DNA double helix winding around each other. Light waves have the same structure, where the two strands are the electric and magnetic fields. We showed mathematically that the relationship between information and entropy can be visualised using just the same geometry.[7]

In an experiment that controlled the transformation of a B-DNA to P-DNA conformation, Jeynes and Parker assessed the energy differences between both forms to verify their theory. The entropy and temperatures can be determined when rearranging the geometry of a DNA molecule from one form to another by pulling its form straight, the P-DNA, (as opposed to being coiled up by nature, the B-DNA) with optical tweezers and twisting it 4,800 turns. Knowing the entropy and temperature of these two DNA versions, the energy differences could be calculated to verify their theory. Jeynes and Parker proved that the golden spiral found within DNA is a maximum entropy state, or the most likely geometry to be adopted by any such system in space-time, in the same way spiral galaxies align as an entropic force to maximize entropy. A system will always go to the state that maximizes entropy. Therefore, the golden spiral, being the uniquity of structure in nature, is a veritable morphic field that emerges through entropic force. In this context, one attribute of DNA is its rapid adaptation to such emergent patterns.

Similar to the approach of the unified manifold of physics, quantum matter fields, geometry, and biochemistry, the morphic field extends beyond our DNA and mind to our surroundings, one matrix to another. In the next section, we will see how DNA is highly telepathic and sensitive to programming—it can detect information from another DNA unit in an entirely different location, reading a thought form and facilitate physical manifestations. Remote viewing, astral projection, and ancestral communication are possible for some due to the quality of their genes' morphic fields.

Quantum DNA

Nobelist Luc Montagnier, known for his study of HIV and AIDS, claims that DNA can be teleported through quantum imprint[8] and proved that DNA emits low-frequency electromagnetic signals that teleport DNA to water molecules.[9] Montagnier's team arranged two non-contacting test tubes, A and B, in close proximity to each other. Tube A contained a fragment of DNA from *Mycoplasma pirum,* a bacterium. Tube B contained pure water. The tubes were surrounded by a copper coil that emits a low-frequency electromagnetic field of seven Hz. The tubes were placed in a box containing mu-metal (a magnetic shielding material), which absorbs low electromagnetic frequency, behaves as a barrier to both tubes, and blocks external sources of frequencies from interfering with the tubes. After sixteen to eighteen hours of sitting in the mu-metal box, scientists used a technique called Polymerase chain reaction, or PCR, to amplify the trace DNA fragments in both tubes. Scientists expected to see only DNA fragments in Tube A. Shockingly, they discovered the same DNA fragments in Tube B. Montagnier repeated the same experiment twelve times with the same results. Russian scientist Peter Gariaev also repeated the experiment and reported similar results.[10]

Wave Genetics

In the last eighty years, scientists have introduced a new branch of science called "wave genetics" that explores the effect of acoustic, electromagnetic, and scalar energy on DNA. Scalar energy, a concept deeply rooted in quantum mechanics and quantum field theory, represents phenomena such as information fields, longitudinal waves, zero-point energy, and quintessence. Research has shown that scalar energy emitted by Tesla-inspired technology can influence E. coli cells, DNA repair, and immune functions. Additionally, Dr. Glen Rein's experiments demonstrated the potential of scalar energy to enhance white blood cell growth, indicating positive influences on the immune and nervous systems. Dr. Victor A. Marcial-Vega further emphasizes the potential benefits of scalar fields in promoting cellular energy and reverse aging while strengthening the connection between universal energy and individual well-being.

How do cells derive consciousness to transmit information encrypted by scalar energy? According to a series of experiments, Gariaev's team indicated that chromosomes emit wave energy and that the electromagnetic fields in DNA are influenced by resonant sound frequencies, having similar mathematical linguistic and entropic-statistic characteristics.[11] Gariaev's focus on quantum holography, a quantum mechanical theory introduced and used by Schempp in 1992,[12] suggests that DNA has a holographic-fractal memory that can be both written and encoded, read and decoded. Gariaev discovered a phenomenon, which he termed "DNA-Phantom-Effect," in one of his experiments: when a laser illuminated a DNA sample from a bull's spleen, DNA converted the beam into a radio frequency, or sound wave. The sound waves influence the polarization of the DNA molecule and create its own lower frequency to transmit information to other genes.[13]

∽०∾

Although my career trajectory was rapidly advancing as I seemed to have created multiple opportunities for myself, I was deeply perplexed emotionally. I became more aware of the emotional atmosphere: the cyclical, monotonous culture, mendacious social treatment, as well as the formulaic scope of the pharmaceutical industry's empirical rituals. In an endless cycle of making the lab's chemical reagents and my project's assays, meeting project deadlines, calibrating laboratory instruments and pipettes, the daily review and upkeep of various project notebooks and experimental folders, addressing quality control and quality assurance findings, attending weekly faculty and client meetings, staying up-to-date with the ever-changing standard operating procedures, drudgingly keeping sample inventories within −20 °C and liquid nitrogen freezers, writing scientific reports, project conclusions, and deviations (it happens to everyone no matter how much they try to intimidate you to be a perfect human, much less a scientist) was the typical eight-hour-plus work day. Not to mention the petty and banal overdramatic orchestrated banter of the laboratory and management—a continual power struggle. Sometimes I debated if I was working in a circus, jungle, or being secretly filmed in a soap opera.

Dispirited, uniform, linear, confined, nonautonomous, careless, artificial, bureaucratic. I felt suffocated. It didn't happen overnight, but each daily dose accumulated to the point where I felt depleted, lost, lethargic, and confused about who I was and what I was setting out to do with my life. At the end of the day, I was another replaceable number to them, treated as and reminded of such.

The deepest encounter that I had with my thoughts about my career was: Was I being true with myself? Was I having fun? Was I going to spend my energy doing something I didn't want to be doing? What did I want? Even though I didn't have the courage to face these questions at the time, I went on to hear them whispered delicately and then ever so slowly and tantalizingly become a crashing crescendo. It was my angels who were watching and were nudging me to sanctuary by getting out.

STORAGE
DNA's Mind-Bending Capacity

The idea of using DNA as a medium for digital information was first proposed in a lecture titled "Plenty of Room at the Bottom" by the Nobel Prize–winning physicist Richard Feynman in 1959.[14] In the lecture, Richard asks, "Why cannot we write the entire 24 volumes of the Encyclopedia Britannica on the head of a pin?" Fast forward sixty years, and in 2018, a team of researchers presented its "DNA Revolution" project at the Sorbonne's Laboratory of Molecular and Cellular Biology in Paris. They had successfully encoded two texts—the Declaration of the Rights of Man and Citizen of 1789 and the Declaration of the Rights of Woman and Citizen of 1791—into the four nucleotides of DNA.[15] The team described the process in this way:

> DNA storage is the transformation of binary numerical data (0 and 1) into letters corresponding to the four nucleotides of DNA. The sequence of nucleotides is synthesized on DNA fragments that can be stored on paper, in a tube, in a metal capsule, and so on. The stored information can then be read using DNA sequencers, similar to those used in biology and medicine for genome sequencing. Once the succession of letters has been obtained, all that remains to be done is to convert it back into binary data, using the same code that was used to write it.

Researchers from the European Bioinformatic Institute reported the accuracy and reproducibility of the information stored within the DNA to be 99.99 percent to 100 percent; their work detailed the storage, retrieval, and reproduction of over five million bits of data.[16]

According to Pierre Crozet, a molecular biologist at the Quantitative Biology Laboratory and the co-founder of Biomemory, one gram of DNA can hold just over 45 zettabytes, or Zo. It would take nearly three million years to download one zettabyte with a fiber optic Internet connection! The human body contains approximately two hundred grams of DNA. With DNA storage, Crozet describes its compact capability as, "all the world's data could fit into the volume of a chocolate bar." As of 2020, the global datasphere reached 59 zettabytes and is projected to grow to more than 180 zettabytes by 2025.[17]

BREAKING THROUGH
DNA Drives Our New Aquarian Consciousness

Emerging into a new age isn't without struggle. Richard Grossinger describes the birth pangs of a coming Aquarian phase of consciousness in *The Return of the Tower of Babel:*

> I treat them as "birth pangs of a new genesis" rather than "death rattles of a dying order" because I believe that everything that has happened on our world since its inception in a stellar spiral—stages of creation described in Genesis 1 and 2, Hesiod's *Theogony*, the Finnish *Kalevala*, Hopi Túwaqachi tales, and countless other ancient and indigenous creation myths (as well as the equally mythical Big Bang of astrophysics)—is part of the universe's evolution from one unknown and incomprehensible phase of knowledge, being, manifestation, awareness, and spiritual freedom, to another. In that sense, it is optimistic and positive, even in its negative aspects, because it is a divine exploration of light and its depth, its immensity, its mystery, and ineluctable transparency and shadows. Because it is light— Buddha's ground luminosity—it will eventually extend in radiance, revelation, and compassion, but not for a long, long, long time.[18]

I turned a corner in the seventh year of my career in pharmaceutical research. After my father's passing, and faced with my deepest and darkest feelings of life, I decided it was time for me to address these thoughts with action. Mornings were early. I woke up at either four or five to begin my triathlon training, shower, and head to work. After work, I resumed my physical training for another two to three hours. My meals included mostly smoothies and juice that I concocted to help my emotional and physical conditioning. I simplified my life. I had to

rebuild from the inside out, beginning by confronting my fears.

I began to seek community, art, and emotional relief through learning to pole dance. I was drawn to understanding my inner erotic woman, who was intensely perplexed and disturbed from being rooted in a sexually rigid marriage interweaved with orthodox religious beliefs. I didn't understand it at the time, but I needed to move my hips, shake my tail feathers, release old energy, and make way for new. It was my refuge. A studio nearby introduced me to a safe space where I could experience a profound sense of sensual rebelliousness, rage, and volition through dancing. It was such a pleasing, foreign, and nourishing environment for me, opening my heart to further dimensions. For the first time in my life, I was experiencing spiritual and hedonist liberties. I went from wearing turtlenecks and long skirts, cautious to reveal any millimeter of skin from the teachings of the church, to managing eight-inch heels, cheeky bottoms, and a loud charisma. It was time to drop my pin elsewhere. I was going to take my peaceful self to a space where I can share this air with others. That was the beginning of my calling.

Stars, worlds, and galaxies spiral around this bright equation before us. Abstracts of ideas that humans furthered through the expression of art, science, trading, cryptocurrency, biochemical monsters, and wars also perpetually make up the chatter and chaos beyond the underlying cosmic thread. How do we position ourselves to differentiate individual genetic development from the current state of confusion the world experiences? We cannot reduce cosmic evolution to chaos simply because we experience it in our world. Genetic evolution is cosmic evolution. What we experience as chaos, or a cosmic bridge, frames genetic enlightenment: it highlights what has always been present and available, morphodynamic memory, telepathic technology, and a storage system through the frequency of our DNA.

In the next chapter, I will show how sound, light, and water demonstrate the qualities of the Rapid Technology Storage system I've presented with my own experiments and frequencies that I discovered using the dodecahedron as a model—the form of our 12-stranded DNA.

Water Memory
and Mysteries

Speaker, Shape-Shifter, and
Heart of the Dodecahedron

Water and sound unravel the secrets of the world.

Water is biological. It is a neat medium that responds to consciousness by rearranging its molecules using the fundamental forces
of nature—strong and weak nuclear forces plus electromagnetism.
In scientific terms, "neat" refers to a pure substance, free of effect or
potency from anything else.

A drop of water, from any source, adopts a number of forms. Its form
shows rhythm, patterns, movement, *and* and inherent intelligence found
in nature. Water's active molecular network is a morphogenetic gradi-
ent for archetypal formation. Its so-called random collection of agitated
molecules is an intuitive network of hydrogen and oxygen particles that
are constantly exchanging information. This interconnected network
could not be governed by the codes of a binary system or any human
algorithm. The intelligence of water is where light and sound, as I have
defined them, create a form that is birthed from the metaphysical realm.

In this chapter, we will explore how water demonstrates the
qualities of the Rapid Technology Storage system, starting with the

profound realization that water itself is conscious and has memory, can communicate, and informs the structure of DNA.

MYSTERIES OF WATER

Aqua consciousness, or water consciousness, is on the fringe of new awareness. Researchers such as Jacques Benveniste, Veda Austin, Masaru Emoto, and Gerald Pollack, have pioneered the field of water in their own ways, showing that water has memory, communicates, and can shape-shift beyond its canonical forms of liquid, solid, and gas.

Jacques Benveniste
Water Has Memory

In 1988, the late French scientist Jacques Benveniste demonstrated that water retains memory through a series of experiments based on successive dilutions of antibodies.[1] His team measured the antibody's activity at full concentration, then measured its activity after multiple serial dilutions. To everyone's amazement, they could measure antibody activity even after there was literally only a single molecule left in the water; then, the activity remained when there was no antibody in the water whatsoever. Benveniste and his team also noted that in order to see this effect, the water solution had to be shaken vigorously before each dilution. In light of all his team observed, Benveniste concluded that water's molecular structure itself could remember and activate the reaction on its own. While the paper, published in the highly reputed scientific journal *Nature*, raised eyebrows, Benveniste's experiment was carefully controlled, following double-blind, double-coded procedures and included six laboratories from four countries. Interestingly, potentization, by vigorous shaking of inert, diluted biochemicals, is how homeopathic remedies are made.

Veda Austin
Water Speaks through Crystallized Hydroglyphs

I've had the pleasure of observing Veda Austin's method, author of *The Secret Intelligence of Water*, one-on-one in a workshop and, on one occasion, was introduced to her team and learned more about their interpretation of the language of water. Her technique, called crystallography molecular photography, produced *hydroglyphs*, which are frozen water crystallization patterns that can be interpreted through symbology. These hydroglyphs inspired my path in merging water consciousness with sound healing.

Gerald Pollack
Water Has a Fourth Phase

The water molecule is special in nature by its three main characteristics: bent, polar, and simple, in terms of representing its fundamental tetrahedral form in nature. These components that make up an individual water molecule create a vast and dynamic interconnected molecular network of magnetic linkages that are constantly exchanging energy among one another. Water acts as both an acid and a base, and autoionizes, behaving like a sun in relation to Earth: where there is light, the opposite side of the world experiences darkness. Traditional hydrology teachers tell students that water has three phases: solid, liquid, and vapor. In 2003, the researcher Gerald Pollack and colleagues identified the existence of a fourth, called the exclusion zone, or simply EZ. They were working with a microplastic called Nafion, a synthetic polymer with ionic properties developed by Dupont in the 1960s, when they noticed that a layer of water could repel the Nafion particles away from a surface. They called this layer the exclusion zone.[2] The same research group continued to study this phenomenon, and by 2007 they used microelectrodes to show that water's "EZ phase" is negatively charged when typically we think of water as neutrally charged, confirming his argument that EZ water represents a separate phase of water.[3] Indeed, the EZ phenomenon is a mystery in science, and researchers propose an array of suggestions for its mechanism—one of them being a change in water's structure. The very existence of the EZ shows how water's molecular organization is advanced and that inanimate nature isn't lifeless. It shows the horizon of a blue revolution in our perception of the world, and we seem to be realizing it now.

Water Changes DNA Geometry

Water's partial positive charges on the hydrogen and negative charge on the oxygen enable its electronegativity to surround DNA in a specific manner. When water content is high, the hydrogen bonds become equally strong and alter the geometry of the DNA strand. The backbone of the helix, where the sugar and phosphate groups are located, bends. The water sheath is a critical part of the DNA structure—even if the aqueous component is increased by just two molecules, the structure folds instantly. As a result, it releases active agents from its dormant or inaccessible regions that are of essential value to our evolution. DNA is an equal partner to water, and vice versa. If water didn't cling to DNA's phosphate groups, we couldn't carry out the foundational expressions of life. Scientists may think of this as algorithmic intelligence under natural selection. I won't split hairs; let's just say, it's intelligent.

The first person to recognize how water changes DNA was none other than its true discoverer, Rosalind Franklin. She identified that the water molecule is what supports the "rich" content, or wet form (both terms simply indicate a higher water content), within DNA fibers.[4] Rosalind's crystallographic calculations and X-ray diffraction imagery highlighted the structural changes that occurred in DNA when its water content varied. A dehydrated DNA form becomes crystalline, shrinks, and is highly ordered, known as A-DNA. A normal physiological DNA structure is hydrated, and its form is longer with a lower degree of order, known as B-DNA (see plate 7).

The original Watson and Crick DNA double helix showed phosphate backbones at the center. Franklin told Watson and Crick that their model contained flaws, because the carbonyl and amine groups of the nitrogen bases could not be measured (by a technique called titration) due to their inaccessibility to water, and she was right. In chemistry, titrations are done to determine the precise concentration of an unknown solution by adding a quantity from a known concentration until a reaction is complete. Franklin's research on water content in DNA crystals indicated dyad symmetry, inferring antiparallel phosphate chains and supporting the presence of two external sugar-phosphate backbones. Her calculations emphasized the volume of water present was critical for its stability.

Now, let's explore how the dance between water and DNA creates the Platonic geometry of their coupling, and how their embrace opens the way to higher dimensions.

THE WATER DANCE
From Tetrahedron to Icosahedron to Dodecahedron

A single water molecule is a tetrahedron; multiple water molecules form icosahedral clusters. The icosahedron is the fifth Platonic solid, containing 20 sides, 12 vertices, and 30 edges, and its face, edge, and vertex correspond to its individual particle spin. The molecular arrangement within an ice crystal is based on tetrahedral geometry. As water melts, the tetrahedron mutates, and the particles that make up its linkages spin countlessly. Hydrogen bonds transition into an ultra-dynamic state of shuffling, or the complex state of breaking and forming.

This water dance is the basis of all life, structure, consciousness and the basis of their archetypal structure.

DODECAHEDRON-ICOSAHEDRON COUPLING
Coupled Archetypes Are a Door
to the Fourth Dimension

The icosahedron and dodecahedron symbolize water and DNA in their tessellated forms. As an inherent characteristic, the dodecahedron interchanges positions with the icosahedron based on the shuffling of the hydrogen bonding within the aqueous cluster, defined as dodecahedron-icosahedron coupling.

Cubes Contain the Golden Ratio

The geometry of a cube—and also many other fivefold symmetries—incorporate the golden ratio, phi or ϕ.* Also known as the golden mean and divine proportion, phi was first described in *Euclid's Elements*. The

---

*The golden mean is found by dividing a line into two parts so that the longer part (*a*) divided by the shorter part (*b*) is equal to the entire length (*a* + *b*) divided by the longer part (*b*): $a/b = a + b/a$ = phi. For example, if a person wanted to create a golden ratio layout in photography, one could set the width to 960 pixels (a = 593 pixels and b = 367 pixels). To print the best resolution prints, 300 pixels per inch is recommended. Therefore, the golden mean ratio can be applied accordingly to size the picture *and* integrate the golden ratio (a = 185.41 pixels and b = 114.59 pixels).

discovery of the constant φ is attributed to the architect and sculptor of Pharthenos, Phiadis (500-432 BCE). The most celebrated attribute of the golden ratio can be found in the Fibonacci sequence.

The golden ratio signifies a numerical value, approximately 1.618, which appears in various aspects of nature, architecture, and sound waves, including music. It is also useful in science in fields such as quantum mechanics, high-energy physics, and cryptography.

Cubes, DNA, and the Fourth Dimension

In the 1850s, Swiss mathematician Ludwig Schläfli introduced what is today referred to as the Schläfli symbol, a notation indicating regular and higher-dimensional analogues of polygons and polyhedra.[5] {n} is a regular n-gon, where n simply represents the number of sides. So, {3} is a triangle, {4} a square, and {12} a dodecagon. The {number}s that meet at a corner add the next dimension. A regular pentagon's Schläfli symbol is {5}, a third-dimensional (or regular Platonic solid) dodecahedron is {5,3}, and an icosahedron would be {3,5}.

A fourth-dimensional dodecahedron, or 120-cell, is {5,3,3}; an icosahedron, or 600-cell, is {3,3,5}. It's interesting to note that one end of the DNA molecule is referred to as 5′ (five prime), and the opposite end is referred to as 3′ (three prime), showing similarity with the designated Schläfli symbols.

Victor Schlegel introduced visualizations of higher-dimension polytopes in 1886.[6] We reviewed the five three-dimensional Platonic solids and the shape of their geometry where each edge is the same length, each face has the same area, and every face has the same number of edges. Polytopes are the "folding" of these shapes into each other in assorted combinations. To better visualize how the shapes look, polytope images use wire projections to depict their "folding." This type of fold represents the bent or warping of space-time in the fourth dimension. Thus, a unique attribute of the fourth dimension is the "folding" of space and, consequently, the ability of one point to jump to another point *infinitely* at any given moment. Because third dimensional (or normal space-time) objects are assembled next to each other, a point transitions from A to B linearly. In the fourth dimension, the bending of space-time creates a path faster than light for two points to be beside one another. For example, the tesseract is a four-dimensional cube folded from eight three dimensional cubes.

I want to emphasize that such diagrams do not represent our natural form and structure, but they *are* our form and structure; that is, Schlegel's topological components make obscure elements and connections visible. Projections of the 120-cell and 600-cell are depicted in plate 8.

Based on these geometries, DNA's coupling with water enables its ascension into the fourth dimension. DNA's nucleotide carbon atoms form tetrahedral structures. Five tetrahedra create a cube. A regular dodecahedron can inscribe five cubes, resulting in a total of ten inscribed tetrahedra. The 120-cell is purely made up of 120-regular dodecahedral cells. The 600-cell and and 120-cell are dual 4D polytopes. In this duality, the vertices of the 600-cell correspond to the dodecahedral cells of the 120-cell, and the edges of the 600-cell correspond to the pentagonal faces of the dodecahedra in the 120-cell. This dual relationship preserves the symmetries of both polytopes, with the structure of the 600-cell reflecting the connectivity and arrangement of cells and faces within the 120-cell.

The fourth dimension extends the simplest abstract aspects of the third dimension—length, width, height—into not just time, but space-time.

SPIN SYMMETRY
A Quantum Magnetics Experiment Reflects How We Can Hear DNA

Now that we have seen how the 600-cell icosahedron interacts with the 120-cell dodecahedron, I believe this relationship reflects an experimental model in quantum research that investigates spin symmetry and its effects on collective energy.

Over the years, scientists and mathematicians have become tremendously interested in subatomic symmetry. Beyond spatial arrangements, symmetry has a value in the quantum world due to its magnetic properties, or the nuclear spins of particles. Metaphorically, spin symbolizes a sphere and its source of life in motion. The benefits of learning about such an exotic material extends to understanding miraculous occurrences such as electrical stimulation without resistance and changes in electrical current in the presence of a magnetic field. Spin symmetry allows us to explore unexpected behavior in quantum matter and cell membrane polarization. After all, the chemistries that underlie biological metabolic processes, such as ATP and GTP hydrolysis, yield highly polar, water-soluble metabolites. Their roles in signal transduction adapt to the spin motion of external stimuli to regulate phospho-transfer activity and coordinate extracellular cooperative binding events.

To research spin symmetry, physicists at the Joint Institute for Laboratory Astrophysics (JILA) created an "atomic clock" by trapping 600 to 3,000 strontium atoms within a laser light. As the atoms collided within the laser, researchers used two electronic levels to track the collisions caused by the atoms' nuclei and examine the interaction between electronic and nuclear spin states that occur in most atoms. Each set of collisions were considered "ticks" of the clock.[7] Strontium atoms are capable of ten possible nuclear spin configurations, influencing the magnetic behavior and random distribution of spin states. Physicists deduced that the electronic-mediated interactions were independent of the atoms' nuclear states—a hallmark of unique symmetry that may unravel interesting collective effects. From one standpoint, Benveniste's water experiments demonstrated nonlocal quantum behavior even in the absence of the antibody's molecular structure. In comparison, JILA's findings demonstrated intrinsic symmetry and stability in electronic spin interactions, unaffected by strontium atoms' nuclear spin configurations. Both experiments laid the groundwork for exploring a molecule's geometric structure, an exotic type of orbital magnetism, and memory.

I believe we can apply a similar model to the collective energy effect of spin symmetry within the icosahedral form of the water molecule. Interestingly, the 600-cell icosahedron model corresponds to the scale of the strontium atoms used in the JILA work, which involved 600 atoms. If laser light can track the nuclear spin configurations of 600 to 3,000 strontium atoms in JILA's experiments, then applying the sound-to-color system proposed in Chapter 10 could enable unique insights when using personally designated color frequencies for each brain-wave oscillation. This fourth-dimension technology could enhance the understanding of DNA dynamics and their connection to brainwave oscillations by visualizing energy states and interactions within the molecular structure.

FOURTH-DIMENSION DNA TECHNOLOGIES

As a technology of the fourth dimension, a designated sound-to-color system that I discuss in chapter 10 would provide an intriguing perspective on DNA's magnetic field using the 120-cell and 600-cell pair as a molecule form. Another model and use of the dodecahedron-icosahedron coupling can be seen in a geographic coordinate system: its longitude and latitude coordinates impose an angular measure within a time period. We can track sunrise and sunset times, elevation angles, and culmination time on such a scale. The dodecahedron-icosahedron dual form can also be seen in adenoviruses. Adenoviruses are common and cause respiratory illnesses such as the common cold, bronchitis, and pneumonia. Adenoviruses possess an icosahedral viral capsid protein, and certain adenovirus serotypes create dodecahedrons during their natural replication process.[8] It is interesting to note that the virus' dodecahedral radius, volume, and coordinates contain the golden ratio, and its placement probably has a lot to teach us.

In the next chapter, we will further explore how sound affects water though the fourth-dimensional geometric approach. The spin is the source of life, and the 12-strand molecule is a self-assembled, super-ionized, sound-spinning molecule: the diva and chanter.

Water Frequencies

Responding to Music, Mantras, and Intention

To build a relationship with water, I began with my goldfish, a watery intimate. I had this goldfish for about a year; it lived in a five-gallon rectangular aquarium next to the window in the kitchen. My name for it was Little Fish. One day, I decided to take a sample of its water and freeze it in a glass petri dish to look at its shape and patterns.* Using just enough aquarium water to cover the bottom of the dish, I experimented to find the sweet spot when the bottom layer of water freezes without the entire dish becoming stressed by ice. I was stunned to see the first crystallized form that the water revealed. There were many odd shapes, but one that stood out was a heart.

I tried it again the next day and continued to talk to Little Fish. The resulting crystallized form appeared to be a lumped flower. The following day the imagery was even more clear, and I saw two flowers beside each other. This time, I decided to consciously talk to the water, and I asked how it would show a friendship. After repeating the same experiment, the crystallography showed two flowers next to each other.

The difference between this and the physics and chemistry of water is intention, not to be taken lightly. In my practice, I discovered the

---

*This was seen using the naked eye; examples are shown in plate 9.

combination of consistency, trust, and belief enchants water: it "shapes water," or "imprints in water." It turns out that these qualities are a means to build a relationship with the molecules. Water will reveal its hidden world when you shift your mental-spiritual orientation to it and acknowledge yourself as its equal and vice versa. When I started my experiments, I expected it to reveal anything I wanted. I didn't treat it as an autonomous being. After a period of time and trials that resulted in empty readings, I decided to shift my perspective. People desire to be treated with respect and acceptance, so I offered this foundational intelligence to water and asked if it would accept me. It did.

One morning, I added two tablespoons of water to a petri dish and introduced myself. It may seem cute and unusual to visualize a woman talking with water, but how else would you treat a friend? "Hi, Water, my name is Ruslana. Welcome to my home. I'm so happy to be accompanied by you. You're safe and surrounded by love. If you sense a chirp, that's my fluff puff cat, Bijou. She's probably asking for food. . . ." It was the start of a vibrant relationship with water's consciousness. I would tell the water what I did that day and ask if it would like to show me how it was feeling, or I would present a question that I had. Most importantly, I was patient with its response. What appeared as random or meaningless squiggles gradually developed into patterns, shapes, and imagery. Right before my eyes, I realized I've been neglecting the source that gives life to all organisms—and that it has its own! It brought me to tears and joy. This type of relationship is what I'm hoping we practice in other areas of life, with animals, angels, crystals, the soil, our ancestors, and of course, our genes.

Mirroring the 2% of Earth's potable water, the same fraction of our DNA is expressed, demonstrates the significant influence small percentages hold over life and individual traits. This concept extends into energy medicine, where practitioners utilize resonances in therapeutic methods to induce transformative change within a person's energy field. In this chapter, we'll explore the scientific pursuit of understanding sound's impact on particles, some early origins of sound healing, and insights from my conversations with water.

VIBRATION
The Action of Sound

Through vibration, sound activates the intrinsic angular momentum of particles. Two different research studies showed this phenomenon. First, researchers at the University of Chicago used X-rays to watch spatial changes in a silicon carbide crystal when using sound waves to "strain buried defects inside it."[1] The study followed a previous work that revealed changes in the spin state of deflected electrons in crystalline matter, which carry quantum information and emit photons. Depending on the state of its spin, electrons influence the activity of spin-dependent photonic emittance. By imaging the sound's footprint going through the crystal, scientists measured the curve of the lattice—a strained molecular arrangement in the crystal lattice caused by sound energy—by tracking a specific point of the distortion in the lattice at a particular point in time. Using a method called Stroboscopic Bragg diffraction, the frequency of the sound wave synchronizes the electron pulses to probe matter at the atomic scale and control semi-conductivity.

Medical research is also incorporating the properties of sound to approach novel healing methods. Outlier cancer research uses sound and music to examine favorable patterns of DNA combinations that would bind to biotherapeutics. The medical researcher Mark Temple at Western Sydney University in Australia remarked, "I realized I wanted to *hear* the sequence. You know, the combination of some sort of audio display and visual display is much more powerful than either in isolation." Temple created his own system of DNA music to find more therapeutic spots in the sequences, which were impressive enough with the initial visual readout.[2] Sound naturally arouses water and DNA compared to many other substances. Water and sound unravel the secrets of the world. Along with Temple, Susumu Ohno, David Deamer, Joël Sternheimer, Linda Long and many scientists have been converting DNA sequences and particle vibrations into musical notes to recognize unseen patterns by *hearing* them since the 1980s.

CYMATICS AND SOUND HEALING

Cymatics provide a valuable tool and a visual aid in using sound as a vibrational medicine. The history of cymatics dates back to the late 1800s, when Ernst Chladni discovered molecular imagery using sound. First, he dusted a layer of sand onto a metal plate followed by a layer of lycopodium powder, and then he played a violin alongside. The finest particles formed into patterns based on the sound frequency. In the 1960s, to build upon Chladni's experiment, Dr. Hans Jenny conducted a series of similar experiments, tagging the phenomenon as "cymatics." During a cymatics experiment, the employment of sound would transform the vibrating medium—using various materials such as quartz sand, lycopodium powder, fluids, and liquid paste—on a metal plate. The pulsation of the form was caused by the excitation of the sound vibration. Now, imagine the idea of a changed form, a "disturbed" form, of sand returning to its original shape. How would one explain such an event?

Jenny believed that our emotions and thought processes are like cymatic forms and are organized by underlying vibrational fields—the densest (the physical) animated by the subtler vibrations, such as emotions and thoughts. Jenny believed that "energy medicine" vibrations are the product of a larger field and a change in the frequency of the field would alter its observation. Jenny ran his cymatics experiments to inspire the imagination toward deeper insights of nature and to demonstrate that energy fields such as emotions and thoughts are transformative as opposed to static.

Organized Vibrational Medicine
Water Replies to My Harp

While conventional science refers to cymatics as "organization," I invented my own "organized vibrational medicine" to explore the behavior of the dodecahedron-icosahedron coupling through calculated sound frequencies tuned to musical notes. The aim of my experiment was to explore the dodecahedron-icosahedron as a natural form of the DNA molecule using water crystallizations and sound. The approach that I used for my experimentation was inspired by alpha-numerology, sound frequencies based on the Platonic form of the 12-strands, and a custom-designed harp.

The setup was straightforward: I added two tablespoons of spring water to a glass petri dish and verbally introduced my experiment to itself. I played my harp with the frequencies outlined in Chapter 17 for 60 seconds, then froze the water in my kitchen freezer for 6 minutes and 55 seconds. I took the petri dish out of the freezer, blotted the back of the petri dish with a dry towel, and took pictures using my iPhone. Any person can access this method (you don't have to own a microscope). Although this approach, which was introduced to me by Veda Austin, is simple and less technical compared to an elegant and scientific setup as such Masaru's, the patterns that emerge are nothing short of magical. The key is to consciously alter your mental orientation towards water (change your frequency) and experiment with the appropriate timing of the water before it completely freezes.

Stars

When I initially observed star crystallizations, my intuitive sense about water was its playfulness. The orderly patterns of the shapes could be effortlessly recognized without the use of a microscope, which validated my original notion that being conscious with the way I relate with water as an equal being is one of the most enthralling and sacred intersections of my life. Similar to a shooting star, the frozen water symbols danced across the petri dish. Some were larger than others and appeared individually, others were surrounded in clusters of stars, and in a few instances—the star appeared from a tight nook just to remind me that it was indeed there. I'm using the descriptor *star* because it is not necessarily a hexagon or pentagon. I've seen different forms of geometric shapes with internal patterning, but the differentiating factor of a star's property is its hollow structure outlined by a *n*-pointed shape without internal patterning. My crystallizations are shown in plate 9.

WATER WISDOM MANTRAS
Water Replies to Spoken Intention

Mantras are a signature sound of meditation that are grounded in intention. Intention, vocally expressed as mantras, have the power to influence water. For instance, in the Hindu tradition, clean water from the ground, *kelebutan*, is a source for drinking water and spiritual transmission among Keramas, Blabutah, and Ginyar communities. Researchers at the Denpasar State Hindu Dharma Institute in Indonesia showed that treating the water using mantras while adding flowers increases the concentration of antioxidants.[3] As a paradox, because the groundwater contains bacteria present from the feces of mammals—coliforms—and other bacteria, the boost of antioxidants naturally sustains all life forms, including coliforms. As a result, there was a natural increase of total bacteria found in the water, showing that even microorganisms benefit from spoken intention.

Dr. Masaru Emoto studied the crystalline patterns of sound's effect on water and revolutionized water research, revealing how water transforms when it is exposed to words, sounds, thoughts, and intentions. For instance, he compared images of water exposed to Mozart's Symphony No. 40, which resulted in a high-resolution image manifesting a hexagon formation, in contrast with water images after listening to heavy metal music, which produced a low resolution form and a rippled sphere.[4]

The following crystallizations were captured from my conversations with water without playing my harp. I wanted to capture the energy exchange through my voice.

> **Stress:** Waking up from a nightmare, the first thing that I did was ask the water about my dream. The red color is a drop of natural dye. I interpret the image shown in plate 10 to depict stress because of the sharpness and aggression of the straight lines.
>
> **Lotus:** I asked water to show what a lotus would look like. In plate 11, the blooming petals can be seen on both sides revealing the flower inside. As a comparison, I include a picture of a lotus with my photograph showing the crystallization of a lotus. The visual aid was not presented to the water at the time the question was asked. The water was simply stimulated by my voice.
>
> **Evolution:** I asked the water if it would show me a projection of the 12-stranded DNA. This "hydrogram" (see plate 12) shows ten straight lines orienting themselves in a single line fashion.

Metaphorically, the strands ascend vertically to symbolize that genetic evolution is a mental and spiritual revolution.

Mermaid: I asked the water what kind of life lives in it. I didn't realize it, but I was also wearing my silver mermaid pendant, which could have something to do with the answer. Crystallized resemblances of what appears to be a mermaid are shown in plate 13. The mermaid seems to be reaching its hand out of the water. I spotted a shape in the bottom left section of the original snapshot that looks like my pendant's fin tail.

As we've explored the fascinating world of water and sound, let's shift our focus to the holistic, multimodal impact of sound within our lives. Johann Sebastian Bach, the "patriarch of musical harmony," created a method to play music in all keys, liberating former musical concepts from their center focus in the key in itself to encompass all of the major and minor keys to create a natural acoustic environment. Not only did Bach's musical genius influence Mozart, Beethoven, Chopin, Brahms and other composers' greatest inspirations, but also inspired current investigations of the relationship between human and music harmony. For instance, it is evident that human perception and physiological signals focus on scale-free signals or natural stimulus by the emergence of brain electrical activity.[5] Bach created a greater oneness that ever existed by developing variations of the musical subject's themes. His contribution extends beyond the musical universe—he introduced harmony similar to Platonic solids and Johannes Kepler's "harmony of spheres." Bach's pursuit to create change, transformation, paradox, and development in music is no different from my own pursuit in introducing the 12-stranded DNA using a range of different keys: the breath, gematria, sound, water, animal and angel guides, dreams, ancestral grief. My composition may differ, but that is the nature of creativity in evolution.

In the next chapter, we will explore how vibrations and frequencies interact with our bodies, impacting hemodynamic, neurological, and musculoskeletal functions, while encouraging the art of letting go for holistic well-being.

Sound Alchemy

The Healing Power of Vibration

Sound emphasizes the subtle, yin aspect of letting go, while activating the memory of our body-mind. It is the most powerful external experience to synchronize DNA.

Ancient healers understood the relationship between geometry and frequency as essential components for creating sacred vibrations. The Hypogeum of Hal Saflieni, a Neolithic underground temple, dates to 3500 BCE in Malta. In this sanctuary, a room called the Oracle Chamber is designed to vibrate at 111 Hz.[1] Inside the Oracle Chamber, you can feel your bones, organs, tissues, and cells at this frequency. This acoustic landscape oscillates disease and addiction out, while attuning listeners to an empathic range. Scientists now know that resonant frequencies shatter cancer cells that hardly show recurrence or metastasis afterward.[2] Yet, this knowledge somehow existed before the times of MRI and CT scans. Our ancestors were the pioneers of sound healing. Even before the existence of humans, the music of the wind blowing through the trees, the rustling of leaves, the waves of the water, and the movement of the grass and sands were Gaia's overture and the beginning sounds that our senses relish and describe as musical notes and terms in modern day. Hazrat Inayat Khan highlights Sufis losing themselves in sound and call it ecstasy, or *masti*, the source of psychic and occult power, and knowledge of the visible and invisible existence is disclosed.[3] In Catholic theology, the Second Vatican Council specifies in detail how music and sound should be employed in the *Musicam Sacram* and *Sacrosanctum Concilium*.[4] The use and culture of drums in neolithic China, ancient Greece, Syria, Sri Lanka, and Africa were used for various forms of well-being, expression, rituals, and ceremony. Further cultures will be examined in the next section.

Sound alchemy can inaugurate repair, activation, and rewriting of the human genetic code because the double helix was built by sound, space, curvature, and energy originally, as was the universe. Sound has a domino effect in aiding the overall functioning of the rest of the human body. Memory module activation, circadian rhythm resetting, creative imagination, and most significantly, the exchange between the subconscious and consciousness—through fantasy or vision—are all examples of sound biogenesis.

The healing application of sound frequency takes many forms, and in this chapter we will look at both spiritual and emotional healing as well as physical changes that therapeutic frequencies empower. Most cultures share the origin of creation begins with a sonorous event. Archeologists suggest the earliest didgeridoo, an instrument used for spiritual practice and mimicking the sounds of nature, was used about 1,500 years ago by the Aboriginal population of Australia to accompany chanting and play an integral role in traditional ceremonies.[5] Ancient Greek physicians used flutes, lyres, and zithers to induce sleep, aid in digestion, and calm the mind.[6] The first published effects of sound in medicine were reported by Diogel in the late 1700s; he played music live by his patients' bedside and proved music lowers blood pressure, increases cardiac output, decreases pulse rate, and supports the parasympathetic system.[7] In Tibetan or Himalayan tradition, rin gongs or suzu gongs are used in mindfulness and spiritual ceremonies by monks.[8] In 1934, Royal Raymond Rife proved viruses and microbes can be inactivated by subjecting them to an oscillating electric field of adequate frequency.[9] Modern science and research show that even animal frequency, such as a cat's low frequency hum (25-50 Hz), stimulates muscles and supports bone healing.[10] Throughout human history, an endless number of accounts in various traditions reference the body's natural response to sound.

In the next few sections, we will explore historical and practical examples that incorporate the effectiveness and integration of sound with human physiology.

VIBRATION AND EMOTIONAL WELLNESS

A practitioner of sound is more than a musician, they are creators. Surrendering to the boundless possibility of each note with intention, she lives in the luminescence of the moment. Laying her hands on her instruments, her being is filled with anticipation and awe for the transformative experience in the offing. She breathes in the vulnerability of space and honors each being, seen and unseen, remembering that each vibration, like each flap of a butterfly's wings, alters the world. Each note is woven within a preceding and subsequent rendition, becoming a tapestry of information warming up a person's internal orchestral framework. Like a gentle breeze flowing through each leaf on a tree, each cell and tissue is subtly revitalized, readjusted, retuned, and realigned from its root to its growth. Even as each note leaves an auditory signature, it continues its vibrational journey within the fibers of our being and enters a new dimension of ethereal creation.

My first sound session was nearly five hours long and changed the trajectory of my life. It helped me recalibrate my voice and awakened courage to leave my ten-year pharmaceutical-research career. I needed detoxification and healing from the combustive mindset and manipulative belief system that generally comes with the corporate technological terrain. Little did the universe forewarn me that I was about to meet the cosmic muse who puts seekers on their paths.

I listened to the healer's heavenly voice as it shifted my attention into a mild trance. The senses of my mind and body transitioned into active tranquility, ready to receive vibrational matter. The first sounds were gentle but clear notes of a metal wind chime. These were *followed by the jiggle of seashells and the sound of bundled chakapa leaves. My emotions began to populate my memories, building from a low resolution into a palpable dreamscape. Sandcastles, seagulls, the slam of the ocean waves, sea foam, and a comforting low wind whistle started to dance in my mind. I didn't feel that I experienced most of these memories in this way in this lifetime. The sound of rainsticks and the soothing melodies of the ocean drum pervaded space. I sensed new emotions with each deep inhale and exhale, attuning my breath and listening. The exploration of my memories were cradled by the harmonics of the 432-Hz Tibetan singing bowl. It brought my focus and awareness of my body to the present moment. At this point, I could feel the shift of my memories become ionized, with love and warmth. They moved to all aura sheaths of my body as I continued pulling nostalgic information that each note unlocked and expressed in imagery.*

When the percussion of the drone from the yidaki began, everything darkened. I slipped into the theta rhythm, an advanced stage of healing that operates within the hippocampus. Located next to the amygdala, associated with stress and emotional arousal, the hippocampus supports memory and learning. Electrical activity in the brain is triggered by neurons that communicate with each other. The synchronized signals become brainwaves. When you meditate or experience attuned sound healing, alpha and theta waves increase. Delta is the first in the dreamscale, theta the second. The theta stage is a captivating zone of consciousness that creates a continuum from the past through the present to the future. It encompasses both precognition and déjà vu central. Neuronal activity linked with the theta stage supports memory-encoding retrieval. Activities in other parts of the brain decrease and allow more focus and energy efficiency such that one can function with less fatigue or exhaustion.

In a sea of memories, I felt connected to the emotions and meanings they activated, even though I didn't experience their physical events within my lifetime in this body. I held each of these memories in love and gratitude. Like a parent watching a child take the first step, speak the first word, or cry from the pain of falling down, I let the sounds carry me through a multidimensional memory lane, a soft channel of tuning forgiveness, patience, and compassion. Fear, guilt and shame, loneliness, not feeling good enough, workaholism, and a marriage with no boundaries, victimization, grief. The fact that these thoughts are far from original makes them no less seminal to the self. They were this lifetime, but karma-driven events radiate through all lifetimes in replicating patterns like ripples from a splash in water.

The next conscious moment that I remembered was a signal to come back into the intimate space as the sound healer summoned us, "And as you return from your sound healing journey. . . ." My eyes abruptly opened. I could feel the nodes of my body fully charged, oxygenated, and hydrated. An uncomfortable knot in my lower back was gone. I had clarity on the next phase of my life. My heart rate reset to a slow and steady rate, sixty beats per minute. Energized and feeling content, I looked down at my watch and saw that it was almost dinnertime. Not only did I spend five hours absorbing a detailed vibration in sound, but I was inspired to learn to become a sound healer in order to facilitate such compelling transformations. I unconsciously accomplished what seemed consciously impossible on an entirely new level—my body-mind "let go," or was subtly realigned by sound.

The Sound of Letting Go

"Letting go" is common parlance used in New Age practices. It is easier said than done. Many of us face challenges in committing to regular practice or fully grasping the phrase's meaning, making it difficult to recognize our personal transformation and truly experience the sensation of surrender. We can't unwind without unwinding our own thread. No one else's thread or method can be a model or stand-in.

Sound healing is, in effect, a specialized type of breathing exercise. It could be described as the yin approach to surrendering in multiple ways. Sound vibrations arouse emotions. Emotions are fluid, like water, having charge and resonance—in theosophical cosmology both are part of the astral realm. Resonance occurs when the vibration of one object affects the vibration of another. For example, a crystal bowl can cause another bowl to join at a similar frequency, even if it's across the room. The "bowls" might be frogs around a pond or crickets in a marsh. When a person is experiencing a strong emotion, his or her heart is affected. Healing sounds, such as the lilt of a harp, synchronize the heartbeat with musical notes, reducing the amount of stress on the cardiac array of tissues. The reggae beat creates a captivating resonance with our heartbeat, matching the tempo of breath and the pulsation of blood. Sound healing offers resonance as one form of letting go to the mind, as I showed in chapter 8 when I discussed sound's influence on water.

Sound healing also induces brain entrainment—a state normalizing brain waves and aligning them to the frequency of a beat. Brain waves generate electrical voltages that allow us to experience the daily world. When we become over- or under-aroused, the fluctuations of our brain waves reflect emotional or behavioral imbalances like those patterns in water. We feel anxious, depressed, or angry, and may experience nightmares. Sound healing can entrain brain waves so that our emotional arousal is stabilized. As we change our perception or attunement, we also change our brain-wave frequency. Brain waves can be altered into infra-low, delta (0.5–4 Hz), theta (4–8 Hz), alpha (8–12 Hz), beta (12–35 Hz), and gamma (>35 Hz) phases. The theta stage, or second gear, is slow and prime during spiritual awareness. It is, as noted, the repository for memories, emotions, and sensations and is the desired brain wave in sound healing for the experience of letting go. In chapter 12, I share a way to shift grief using the sound of your voice as brain entrainment.

BIOGENESIS

When we listen to sound in meditation, the brain switches off the language center within the prefrontal cortex, allowing for enhanced creativity and imagination as it induces a new phase at the given frequency. When cell-based DNA receives sound, it doesn't constitute a specific "treatment" like conventional medicine. The entire body is consumed with vibrational resonance and is recalibrated. Sound waves have been shown to interact with cellular structures, and there is ongoing research exploring the potential for sound frequencies to influence DNA stability. In effect, sound healing treats sickness from the body by an emotional mechanism.

Sound healing and meditation can also lead to neurogenesis. Scientist Dr. Graham Phillips did an eight-week meditation experiment to track the structural and functional changes in his hippocampus. With MRI analysis before and after, he showed that the bilateral hippocampus volume increased by 23 percent, particularly at the dentate gyrus, where new brain cells are produced in adults.[11] Evidently, these changes are uncommon in older adults. Although young people generally show this pattern, Dr. Phillips proved that meditation or sound healing reverses the aging process.

The application of utilizing sound frequency as a therapeutic modality is one of the few forms that calibrates the left and right hemispheres of the brain equally. Although science divides brain function into different hemispheres and functions, sound treats it as a whole brain. The state of the brain influences the growth and repair of cells, tissues, organs, bones, and the biochemical systems of the body, and programs instructions of physiological surrender. The health benefits of sound and mechanical vibration were recognized from carriage rides on cobblestones in the eighteenth century. Abbé de St. Pierre developed the *tré moissoir* or *fauteuil de poste*—a vibrating chair—to soothe melancholia, liver disease, and Parkinson's disease symptoms.[12] In 1878, the physicians M. Vigoroux and Jean-Martin Charcot used a sounding box with a large attached tuning fork to treat patients with migraines, hemianesthesia, and locomotor ataxia.[13] To further investigate the idea of mechanical vibration on the brain and inspired by sound therapy, Gilles de la Tourette (Charcot's assistant) invented a helmet with a motor on top that caused the helmet to vibrate at 10 Hz—he found it had a positive effect on insomnia, depression, and migraines along with other conditions.[14]

In 1996, the Federal Drug Administration (FDA) approved a treatment called vibro-acoustic therapy, officially called the Next Wave Physioacoustic Chair. The therapy involves six imbedded subwoofer speakers that apply low frequency sound vibration—frequencies from 20 to 130 Hz—from the knees to the shoulders to increase blood circulation, decrease pain, and increase mobility.[15] Sound vibration results in compression and decompression of air molecules. The actuator travels to the receiving mechanoreceptors in the skin of the body and to the ear's tympanic membrane. The resulting compression and decompression from the surface of the body is delivered to the bone and tissue.

In light of these therapies, let's explore mechanisms activated by sound vibration on the human body: hemodynamic, neurological, and musculoskeletal.

Hemodynamics
Vibration Affects Blood Circulation

Hemodynamics measures blood flow, including arterial pressure or cardiac output. The physical study of hemodynamics is essential to understanding the functioning of the circulatory system. Nitric oxide modulates blood flow and oxygenation of tissues. Vibration stimulates endothelial cells and lymphatic vessels to release nitric oxide and induce blood flow by mechanoreceptors, although the mechanism for this sound activation is unknown at this point. The FDA also confirmed the increase in blood circulation by vibrational effect.[16]

The activity of osteoblasts and osteoclasts maintain the extracellular matrix of a bone structure. Osteoblasts synthesize and materialize the bone matrix, and osteoclasts degrade bone to mediate normal bone remodeling and regeneration due to bone loss from pathological conditions. Sound vibration, specifically 40 Hz (the sweet spot of a cat's purr), promotes the activity of osteoblasts by upregulating RUNX2, a genetic transcription factor associated with osteoblast differentiation.[17] Even with mechanical vibration at 30 Hz, the expression of RUNX2 increases.[18]

Sound vibration can also reduce blood clotting. The mechanical action was seen in a study that dissolved a pulsating blood clot by 25 percent using 50 Hz vibration for twenty minutes.[19]

Neurological

A fetus starts to hear the rhythm of its mother's voice and heartbeat at about twenty-four weeks, as neurons form connections and migrate in the auditory cortex.[20] Low-frequency sound vibration appears to stimulate neurite outgrowth and neuronal differentiation. In a study that examined protein expression related to neural differentiation, 40 Hz sonic vibration stimulated protein calponin 3, which promotes neural differentiation.[21]

The proprioceptor system includes receptors, nerves, spinal cord, and pathways of the central nervous system and influences the control of posture and movement.[22] Vibration's effect on the proprioceptor system has been a widely researched topic in neuroscience as its sensitivity plays such a dynamic role in motor control and neurological diseases. Cerebral palsy,[23] multiple sclerosis,[24] and chronic musculoskeletal pain[25] have all shown a positive clinical effect when using vibration treatment, including one study that found significant improvement in balance and motor skills in a randomized trial of twenty-four children with cerebral palsy.[26]

Pain is a complex concept to understand because it is not clear what represents pain. Not only is pain a personal and subjective experience depending on the individual, but sometimes pain can exist without any known cause. Regardless, vibration can act as an analgesic to pain. Numerous research studies have shown that vibration can reduce pain in orthopedics and low back pain,[27] physiotherapy,[28] and cosmetic procedures.[29]

Musculoskeletal

The musculoskeletal system is a natural vibrational cascade of events: neurons stimulate motor neurons, synchronizing their frequency to muscular contractions and increasing muscle fibers and protein synthesis, leading to muscular strength.

In rehabilitation sciences, vibration can improve the tone of muscles or act on their behalf. In studies performed by Gloeckl and colleagues[30] and Lage and colleagues,[31] vibration therapy helped patients with chronic obstructive pulmonary disease (COPD). The patients needed physical exercise but were unable to work out because of shortness of breath. Also, extensive exercise can stimulate pro-inflammatory cytokines that inhibit the muscles. In these COPD patients, vibration therapy supported mobility and influenced an anti-inflammatory effect. These therapies are also beneficial for frail, elderly individuals experiencing sarcopenia, a form of muscle loss due to aging or immobility.

Although these are only a few plots on the map, the vibrational effect of sound on the body is shown in numerous research studies. I would like to emphasize that the way vibration at different frequencies affects the blood, brain, and bone is complex. Clinical, often reductionist, research assess the effect of vibration as binary, as either positive or negative, but this dualistic system can be deceiving for many reasons, one being that an integrated field of vibrational research does not exist.

How *do* we interact with variables across multiple mechanisms such as frequency, cycle, patient group, ranging applications, and amplitude? Furthermore, sound vibration therapy on human health is an exploratory, traditionary, and traditional approach. If we can inquire into the stimuli of sound internally, I believe our curiosity will lead to our own particular medicine bouquet of healing.

PART 2

BREATH, ANIMALS,
AND ANGELS

*Heal, Activate, and Evolve
Your Dodecahedral DNA*

Introduction to Part 2

What are the healing processes? The word *healing* can be misunderstood as a crutch or a failing, as if something is wrong or in need of remediation. Yet the ordinary pattern of life is healing. Cells organically repair and replicate themselves. By their ontogenetic nature, they differentiate, divide, die, replicate, and continue to evolve. This sort of healing is separate from a hospital, church, clinic, or laboratory. It is the activating state of evolution.

It's impossible to *not* evolve. In the context of this book, healing is simply a redirection to work with pain. Sometimes pain is perceived as a punishment. However, pain is a language of healing, showing a detour or an undertone to try something different. Without accepting that healing is an organic process of evolution, we become static, complacent and unbalanced.

It is understandable that people often equate healing for correcting an error or repairing a psychological flaw. How many times have you experienced unsolicited advice from people who want to help "fix" your, or their, life? How often have you been offered how-to guides for life improvement? When positive changes don't last, despite an individual's best efforts in detox programs or other self-help endeavors, it can be disheartening. However, it doesn't imply failure or the impossibility of change. Healing goes beyond the mere absence of toxins. Similarly, the absence of war doesn't mean the presence of peace.

My approach to healing with 12-stranded evolution is that there is nothing to fix. The interior coding will actually regulate fear, anxiety, anger, and your previous thoughts about life and death, aligning them with the rest of the universe. My intention is to interweave submerged memory in order to cultivate wholeness, agency, and justice, while dismantling, refining, and expanding upon familiar images that may hinder our understanding of healing. My application to well-being supports the silhouette as well as its radiance.

Medical intuitive Laura Aversano wrote in "The Nature of Evil," an unpublished channeled text, "With some understanding of human relationship and the potential for it to serve mankind, we strive to understand the nature of evil, the responsibility of each participant, and the responsibility of God." In seeking a higher purpose for our shadow, we can begin to respect the thoughts and emotions that birth the creation within. Subsequently, our memory strands are activated and spur our evolution. In part 2 of this book, we will explore breathwork, embracing and shifting grief, animal energies, and our inner angelic ferocity to activate our DNA and evolve to our highest selves, taking our place in the cosmic, collective consciousness.

Draw Breath and DNA Consciousness

The breath encompasses a realm of endless possibilities and experimentation.

Breath is rooted beyond the threshold of mortality, encompassing a realm of endless possibilities and experimentation. It has the power to create space in our body to regulate emotional disarray such as depression, mood changes, anxiety, mental focus, pain, and grief. When treated with intention, we can align our energy complexes so that the body-mind can perform as a harmonious system. The breath connects our mind to the body so that we *feel* our existence, rather than coast through it.

We tend to overlook breath's capabilities by attaching its function to survival—seeing it as a plain, unconscious, and vanilla mechanism. In 2021, Croatian diver Budimir Šobat held his breath underwater for 24 minutes and 37 seconds,[1] showing the world that the relationship of breath, body, and mind is more intricate than initially believed.

Breath moves conscious intelligence from your lungs, circulating to the heart, neuromuscular system, and your cells, to your DNA. As we move our breath mindfully, we direct this intelligence toward our cellular intelligence. If our breath is sluggish and deathlike, our life translates into a self-sabotaging illusion. If our breath encompasses imagination and compassion, we become inseparable from life and knowledge. If we practice moving forgiveness to our cells with the vehicle of our breath, we begin practicing sacred healing.

I developed a business plan during my last year in the corporate sector. Inspired by my smoothies and wanting to share the medicine of healthy eats with the community, I began my entrepreneurial venture with a food truck. It was quite a jump from being a scientist to a smoothie-slinging barista in a fifteen-year-old box truck. Did I have any experience in the food industry? Zilch. Did I understand the technicalities of running a mobile business? Most definitely not. What type of support system did I have that I could lean on when I wasn't sure what to do next? I had my breath, hope, and a dream.

The dream started with a vision of a Key West shade of blue used box truck. Seeing a listing on Craigslist, I met its owner, Dean, and my future restaurant: an '88 Ford, with 200,000 miles, and—most of the time—it would start. Dean and I conversed for about twenty minutes, and shortly after, I handed him a wad of cash. While he counted, he asked if I even wanted to see the generator. Staring at my truck as if I just met my future dreamy husband, I said, "The generator? Oh, yes! That! Absolutely." It turned out that the truck generator died at my first event, and I had to replace it. My Key West-blue gem became a unique hit in my hometown, Richmond. It was the first truck that served smoothies and juice, under my alias, Pulp.

By reprogramming the way I breathed, I changed my aetheric language. Small shifts each day became noticeable in the way I handled mishaps and oppositions. I led by breath and free will, away from "you have to," and flowed with life.

I practiced this form of breath freedom (and bravery, for me) which translated to spiritual power over the next seven years in all of my endeavors. I experimented with different identities and measured up to them: a scientist, triathlete, a dancer, a food truck operator, a coffee shop owner, a business consultant, a podcast host, a videographer, and a writer. The breath taught me sound and light medicine. It inspired the creation of my own music and sound healing experiences. Breath leads my DNA meditations to help people activate their 12-stranded DNA.

DNA BREATHES AND ACTIVATES

DNA has its own breath, even as atoms have their own gravity. When DNA breathes—a process referred to as "fraying"—the strands spontaneously fluctuate in shape. At physiological temperatures, DNA's structure is stabilized; however, occasionally, with small temperature fluctuations, the base pairs can decouple, or fray, opening and exposing the otherwise buried bases located within the DNA duplex—in other words, "breathing." These DNA breaths initiate a multitude of genetic processes. DNA is a precursor and reflection of our humanity, sharing not only breath but also memory and consciousness.

We can see that DNA has inherent memory in several ways. First, the DNA molecule gives rise to a variety of life forms—viruses, proteins, and enzymes—that originate from the realm of molecular consciousness. DNA naturally regulates and modifies other genes; for example, DNA methylation is its way of keeping a gene from being expressed; it is effectively turned off by the methyl groups binding and covering the transcription start site. DNA can interact with other forms of itself such as mitochondrial DNA and viruses. It also "knows" how to repair itself by interacting with its external environment to induce change. All of these activities are governed by DNA's evolutionary memory and inherent consciousness.

I believe the matrix that holds these multiple interactions between these consciousness states is the breath. The breath transcends the notion that consciousness can be reduced to a purely mechanistic model. The extended mental field resulting from this interaction-based model between DNA and breath consciousness is the *cosmic breath.*

THE COSMIC BREATH

Breath is the innate pulse of creation and how plants and animals—even bacteria—orient themselves in an animated rhythm. These rhythms manifest in dreams, visions, animals, angels, sounds, oracle cards, numbers, mediumships, and synchronicities, shaping our understanding of magic and divination. But even these examples are fragmented within their own paradigm and dualism. Such is the tug and pull of the cosmic breath as it whispers its wisdom—as the soul's reminder of who and what it is.

Bursting into Being
Soul Consciousness Begins with New Life

The cosmic breath activates soul consciousness at the very moment a sperm fertilizes an egg, forming a zygote. In *Your Spark Is Light: The Quantum Mechanics of Human Creation*, Courtney Hunt talks biology as if it were physics:

> At the moment of the collision of two Higgs fields of the sperm and egg, they create a microscopic black hole. The collision of these Higgs fields generates enough energy to create a new Higgs field that is trapped by the 20 billion zinc atoms released . . . delivering the soul consciousness or extensive ZIP code if you will to the newly formed zygote.[2]

She is depicting energy, fields, information, and vibration—cosmic breath—translating between tiers of microcosms. Black holes are incredibly dense formations in space with strong gravitational pulls, trapping everything within, even light. In another study performed at Northwestern Medicine, scientists observed "zinc fireworks" through fluorescence microscopy when an egg was activated by a sperm and became a zygote.[3] It can be represented as something like the activation of Dr. Hunt's microscopic, subatomic, black hole.

Cosmic Melody as the Soul's Gateway

Sound, produced with the cosmic breath, is the rhythm of our minds and the gateway to our souls. I believe the origin of the soul can be mapped to brain oscillations as a function of chromesthesia—a type of synesthesia in which a frequency or sound evokes color. All music is created using the infinite combination of the notes E-A-D-G-B-E. When we designate a color to each note and assign those to brain waves, we can visualize the movement of subtle energies with a vivid lens. Sound transforms into color or thoughts, making our world an animated tapestry of frequency, amplitude, and source.

Russian composer Alexander Scriabin believed integrating colored light with symphonic sound would be a powerful psychological resonator.4 In 1911, he published a sound-to-color map: E (blue); A (green); D (yellow); G (orange-pink); B (similar to E).5 However, individuals can create their own unique sound-to-color associations based on personal experiences and perceptions.

In my personal exploration of the relationship between light and the cosmic breath, I created my own color designations for each brainwave oscillation frequency.[6]

Delta, < 4 Hz: E—Blue
Theta, 4–8 Hz: A—Green
Alpha, 8–12 Hz: D—Yellow
Beta, 12–35 Hz: G—Orange
Gamma, > 35 Hz: B—Red

I propose that a zygote, or fertilized egg, contains a pitch of light, which serves as a portal or gateway for the soul's incarnation. The energetic crystal of the soul, as simultaneously dense and transparent as the blue sky, integrates with the physical realm by transferring its aetheric DNA into biochemical DNA. These nucleic acids line up the way they are supposed to, like beaded string trinkets or soldiers in a military drill. I don't know the exact frequency at which conception takes place, but gamma is the highest processing frequency of the brain and marks the first wave of fertilization and soul attraction, followed by beta, alpha, theta, and delta. This cosmic melody shows a trans-physical synapse between the soul and fertilized egg, where the soul raises its pitch (represented by musical note B) and is developed as a gene-based living organism.

The burst of zinc at fertilization depicts a harmonic convergence of two organisms serenading to form a new life, imbued with soul consciousness and defined by color, frequency, and rhythmic information. This synesthetic model offers a unique lens to explore the aetheric matrix that surrounds us, the cosmic breath.

THE GEMATRIA OF NEW LIFE

The Greek word for "zygote" is ζυγωτός and "soul" is ψυχη with corresponding gematria values of 1780 and 1708, respectively. Their values not only share number similarities but are also separated by the number 72. Seventy-two is a widely known number in angelology, theology, and spiritual traditions.* Also, it pertains to the number 12 as regular solids—the dodecahedron and icosahedron—and their forms in the fourth dimension: the 600-cell (720 edges) and the 120-cell (720 faces). We looked at these closely in chapter 7. Zinc is a member of the twelfth group of the periodic table, and its atomic number is 30, which is also the number of edges on a dodecahedron and icosahedron.

Gematria can also be used as a tool to reveal knowledge as a temporal marker. Zinc or ψευδάργυρος contains a gematria value of 1983. In 1983, biotech saw several significant events: Kary Mullis discovered polymerase chain reaction (PCR), Montagnier and a team of scientists discovered the human immunodeficiency virus (HIV) as a possible cause of acquired immunodeficiency syndrome (AIDS), and the Internet was invented. The world was activated by going online, similar to the way zinc sparked the zygote. In the same way I compare zinc to historical events, another comparison can be seen with the gematria of Adolph Hitler, Αδολφος Χιτλερ—1920. In 1920, he gave his speech and initiated the Nazi Platform, rising to power.

Now that we've explored the elements of the cosmic breath, let's move on to how to apply our healing DNA breathwork: a methodical breathing technique to prime the state of the body for psychic programming

---

*There are numerous traditions that reference 72. The number 72 is widely seen in the New and Old Testament, Buddhism, Coadaism, Egyptian mythology, Chinese perspective, and so on. Particularly, chapter 14 will address the number 72 as it pertains to the Tetragrammaton and its relevance to Plato, Fibonacci, Arcturus, morphogenesis and the new sound frequencies discussed in chapter 17.

and genome activation. The primary focus of this breathwork involves using movements as a means of replenishment for DNA, activation for memory, and memory storage and retention.

Traditional education often overlooks the importance of teaching proper breathing techniques and the significance of conscious breath for the mind and body. Most of us exist without awareness of the unconscious movement of the breath. It affects the way we think about ourselves and the world. When you receive a massage, your body gets worked on to remove knots and sores that you are unaware of. You leave feeling relaxed and calm, enjoying the feeling of revival. Similar types of "knots" form in our physical and aetheric body when we don't breathe consciously. They form rigidity in our body-mind, leaving us feeling impatient and irritable.

When I was a scientist, I wasn't interested in prayers, telepathy, or psychic healing. Being a rational, empirical professional, I thought it was nonsense, a fairy tale, and a waste of time. If a breath, thought, or prayer affects things in the physical world, its influences should be measurable, and science should be able to investigate them. After taking a step back from biotech and corporate policy into holism, I understood there is more to nature's complexity than what we understand from chemistry, physics, or synthetic biology. My personal view is that there are other causal factors in nature that bring about new kinds of knowledge, such as the breath, thoughts, and mantras for healing.

If you want to read someone who has arrived at this perception by a different path, check out Larry Dossey's Healing Words. *He reviews 131 controlled experiments on prayer-based healing, and more than half of them showed improvement in the person being prayed for.[7] In a double-blind study of 393 patients in the coronary unit at San Francisco General Hospital, 192 patients were chosen at random and prayed for by formal home prayer groups, while the others were not. The patients that were prayed for recovered better than the controls, and fewer died.*

As a former scientist, I recognize multiple scientific and theological implications to such a study. I could ask a hundred unsettling questions that expose inconsistencies in both science and faith, such is the nature of scientific critique, comparison, and human parallel. Proving that prayer cooperates, or studying its efficacy, is an independent confounding variable. Therefore, it is impossible to draw a conclusion from any randomized controlled study on prayer based on convincing results, because they're two separate models. My perspective of a spiritual healing experience is based upon faith. In my practice, my medicine comes from my belief, and it works.

DIFFERENT APPROACHES TO BREATHWORK

Breath isn't limited to one form. The brain adapts our breathing patterns according to the needs of the body. Among multiple breathing techniques known, many are familiar: asana, pranayama, box breathing, valved breathing, nostril breathing, vocal breathing, and the breath of fire used commonly in Kundalini yoga. These diverse breathing arrangements, though varied in their origins and applications, have proven effective in harnessing calmness and clarity.

The Wim Hof Method

Established in 2011, The Wim Hof Method (WHM) incorporates cold-water exposure, a variety of breathing techniques, and willpower.[8] I believe essences of the Wim Hof* Method are vital to the cellular activation in this book.

The Wim Hof Method in Practice

Assuming a comfortable meditation posture either sitting or lying down, inhale forcefully through the belly, chest, and mouth and exhale through the mouth repeatedly thirty times in short powerful bursts. After the

---

*Wim Hof is a Dutch extreme athlete and educator. Hof has broken several records in the Guinness World Records: In 2013, Hof held the world record for the longest full body duration with ice for one hour, 53 minutes, and 2 seconds; in 2000, Hof claimed the longest swim under ice while holding his breath at a distance of 57.5 meters. To this day, he holds the record for the fastest half marathon barefoot on ice and snow at 2 hours, 16 minutes, and 34 seconds. Hof founded a psychophysiological method that aims to reduce stress responses and hormone release.

thirtieth inhale, take one long inhale and exhale, letting the air out. Hold until you feel that you need to breathe again. For the recovery breath, draw one deep inhale to fill your lungs. Hold the breath for about fifteen seconds, and let it go. Repeat the cycle three to four times. In rare cases, it can cause a loss of consciousness or impact motor control due to the body's sensitivity toward this controlled method of hyperventilation. The method requires stillness of the body so that the mind is present in each cell.

Cold Exposure

Inspired by the Iceman, Wim Hof, and his method of cold plunging, my partner and I set up an ice bath at our home using a chest freezer from my former brick-and-mortar coffee shop. Cold plunging is a term for therapy by systemically surprising your body with cold water (10°C to 15°C) in any shape or form such as swim, shower, ice bath, or even a frozen ice pack—while breathing calmly. Deliberate cold exposure is an age-old tradition to improve metabolism and quality of sleep, calm the nerves, dissolve depression, and lead to athletic recovery and general well-being. With practice, I learned to sit calmly in freezing water with heat that I could feel radiating from my heart for at least ten minutes. The experience helped me stay present in the moment, keeping unnecessary thoughts at bay. You could even find a smile on my face.

Another benefit of the Wim Hof method is the blood level's pH value increases, becoming more alkaline. With less acidic blood due to the fluctuation of carbon dioxide to oxygen ratio, the body produces more ATP (adenosine triphosphate—the source of energy that drives and stores cellular processes) more efficiently while inhibiting lactic acids. Hof even proved his breathing and cold plunging therapy influences his autonomic nervous system by participating in a study where he and other participants were injected with *E. coli* endotoxin.[9] The anticipated symptoms were common flu-like symptoms, headache, fever, muscle pain, nausea. Yet, Hof only experienced a mild headache, showing that he produced only half the number of inflammatory proteins compared to the other participants. The explanation for Hof's controlled response to the bacterium lies in his concentration method.

Let's begin by using a basic breath practice.

Draw your attention to the rise and fall of your chest at the same rhythm as your nasal cavity. Sense the synchronism. Feel the breath entering your body and leaving. Relax your belly and allow it to completely hang with no resistance. My vocal mentor calls it the "loose belly." This is the physical state of your belly when you inhale. When you inhale, take a deep breath to expand your belly.

On the exhale, imagine your belly button drawing back into your body. Making a hissing *S* sound using your mouth until you can't make the sound anymore while you draw in your belly can help you understand the motion. It will sound like a deflating balloon. Take your time and fully allow your body to take your belly button as far as it will go. When all of the air has been exhaled and you feel that your belly can't move further back, inhale and repeat.

Eventually, it becomes how you breathe.

BREATHWORK PRACTICE TO ACTIVATE YOUR 12-STRANDED DNA

This breathwork method uses a stabilized approach to hyperoxia, excessive oxygen supply, and hypoxia, reduced oxygen supply for DNA transmutation. I recommend listening to music to support your mind with rhythm and allow your breath to access the deeper layers of genetic fascia. Our mind is constantly guarding our emotions to protect the ego, so we have to trick it (I call it coyote medicine) to mitigate these boundaries through emotional arousal in order to heal. Our feelings have a way of melting such barriers, especially when we're listening to music that touches our heart. Be comfortable in this breathwork; most people lie down on their back. People with lower back complications can sit upright in a seated position.

Anyone who has cardiovascular constraints should be cautious doing this breathwork. Ask your doctor if your heart is strong enough for any type of breathwork, especially this one, and what techniques would be best for you. If you're on medications, have a history of seizures, or had a recent injury or surgery, practicing breathwork could pose potential risks. Pregnant or breastfeeding women should not do this breathwork.

Basic Guidance

Begin by lying down or sitting in a comfortable upright position. Take a deep inhale through your nostrils and exhale through your mouth with force. Inhale assertively (i.e., no longer than 3 seconds) and exhale through the mouth vigorously (also between 2 to 3 seconds). Relax your jawline and allow yourself to be completely vulnerable to the exhale. Feel that you can let it go. You can allow natural sounds, such as *haaaa*, in any shape, form, or amplification to escape on the exhale. The *vuuuu* of Peter Levine's* somatic experience is a powerful cleansing tool in raising your vibration.

Do a set of 15–20 of the indicated breathing. Then hold your breath for 30 seconds. Shift to slower and longer breaths for 2 minutes. You can discriminate reps by duration of songs to track how much breathwork you've done. One set is approximately one average-length song, three minutes. Repeat for three to four songs.

Because this is a multitiered method, I've enlarged the focus to encompass each level, physical, mental, and spiritual. These factors will differentiate your experience, which is the intention of this practice.

Physical Guidance

When you inhale, stretch your abdomen and expand your diaphragm. When you exhale, release your diaphragm, moving the breath down to the abdomen to your sacrum and coccyx, or tailbone. The goal is to end at the tailbone before you take your next inhale. You may find your tailbone will gently lift off the ground and tap into the ground on the exhale as a marker of this style of breathing. Don't let the tap be your overall focus, simply keep it as a physical destination of the breath. This is a full revolution of the breath, from nostril to your root and root to nostril.

---

*Peter Levine is a teacher and psychotherapist renowned for his research in trauma therapy.

Mental Guidance

Visualizations are helpful to keep your thoughts organized and in sync with your breathwork. Imagine a cord from your brain to your root. The brain is the body's electrochemical organ that connects and processes information from the cosmos and mind. The tailbone is the body's charging station to the ground, a neutralizing point of information transfer with the earth. See a holy electrical current, facilitated by the breath and beginning with your mind sending oxygen to your lungs, supporting the heart's performance. The mind programs oxygen into intelligence in the form of love, gratitude, forgiveness, acceptance, trust, and joy. Let your imagination decide on the shape or form.

Some mantras to pair with the breath are, "I feel love. I am safe. I am grateful. I trust my breath and evolution. I send these vibrations of intention to my DNA to evolve. I open the channel of genetic memory. I am present in each cell." Don't try to compartmentalize or concretize each part of your channel, this is general flow. The equipped oxygen circulates to your abdomen, sacrum, legs, feet, and toes and replenishes each cell with love, which is merely a name for the energy. I could also call it elixir or alchemical solvent, but "love" connects it to the heart. Once the charge of the current is neutralized by the ground, you've created an organized, rotational field of healing through breath.

As you adopt your own rhythm and pace, imagine you are pumping oxygen into your body, and redirecting its course using the mind as heat, like a hot-air balloon. Your body will even start to feel a lift off the ground. You will start to feel like you can even walk on water. Stay calm. You are experiencing a new level of consciousness. If you want to regain ground, close your eyes and take slower, longer inhales and exhales. You will begin to feel earthed again.

Spiritual Guidance

Imagine you are gathering oxygenated nectar with someone you admire. When you meet them, greet them with welcome and warmth, perhaps a smile. When we smile, our body and mind become airy and lighter. We take in more information when we relax. A small thing, but when we laugh, our body allows the information to be digested and metabolized with ease, removing its formal edge.

When we perform this breathwork astutely, smiling may be the last thing on your mind. However, when we make ourselves smile in the process, we are also focused and glorify the experience further by opening our channel more widely. Breathe with your full heart and mind, as if you're getting ready to meet your future evolved self, a cosmic cousin, or a new revelation, because you are. Try to smile in the process, especially when you don't feel like it. Be pleasantly surprised with the tremendous difference in the ball of energy you'll unravel and feel, as against a tightly wound-up mass of material.

In cultivating this breath, we can reprogram it and start to witness our DNA awakening in the infancy of a new self. You can move toward the DNA editing capacity of our invertebrate rivals in the cephalopod kingdom. We both start with the same blastula, then play different DNA chords and breathe different DNA stops and channels. Some osteopaths even "feel" a new bone starting to form at the dens or odontoid process near the base of the human skull. There is no bone in the X-ray, but they sense its looming aetheric presence.

Cool Down

After doing a three- to four-song repetition, you will feel your body's energy transform into a state of excitation. When you reach this peak in your breathwork, your mind, body, and soul are activated. You will feel energy fields in your hands, feet, arms, legs, throat, sacrum, abdomen, pineal gland, heart, and root. When you're ready to move on, cool down. The cooldown is equally important because it integrates the energy within your body. Similar to baking bread, kneading the dough is critical in allowing the protein molecules in the flour to form, creating healthy gluten strands. In this case, it is your DNA strands.

Starting with your lower limbs, one foot at a time, slowly move them around an imaginary circular axis beneath your ankles until you are satisfied. Move up to your pelvic bone, take each leg separately and gear them in a circular motion, taking your time. You can use your hands to help. Repeat the same for the sacrum, abdomen, and chest. Try to feel any new bones in your psychic field. When you get to your head and throat, slowly roll your head in a clockwise and counterclockwise direction. For your skull, visualize a spiral forming from the center of your forehead, in the middle of your brow. Energetically it forms an upward spiral and dissolves like incense into the air. All of these steps are meant to be gentle and at your own pace. Listen to your body through your breath. Learn to notice if you feel your breath feels spaced, toned, and sensitive to both mind and body. Your system is shifting to an alkaline state.

You've completed the 12-stranded DNA breathing method.

Considerations and Important Questions

Is it dangerous to make myself hyperventilate?

Hyperventilation itself is not a danger. The greater danger is either shallow breathing or breathing uncontrollably for long periods of time, which contributes to the drop of carbon dioxide in the blood. Carbon dioxide functions as a vasodilator, keeping the veins open for blood regulation. When you hyperventilate, the carbon dioxide drops, and the veins contract, decreasing the supply of oxygen to muscles and organs. Controlled hyperventilation consists of lower reps, allowing your body to stabilize its oxygen supply. Also, go at your own pace. The focus of this practice is lost if you hyperventilate. The euphoria of DNA breathing is not lightheaded; it is deep-headed. The objective is to achieve a sensitive level of consciousness so that you can become attuned to your subtle body through the breath.

Is it normal to feel my muscles lock up?

In this breathwork, there are three common experiences: muscle spasms, tingling or nerve twitches, and the aforementioned lightheadedness. In scientific terms, the body experiences a decrease in carbon dioxide partial pressure (the measure of carbon dioxide in the blood), an increase in blood pH (resulting in a higher hemoglobin affinity for oxygen), and the blood is less likely to release oxygen to the muscles and tissues that need it the most. Conversely, as carbon dioxide increases, the blood pH lowers and decreases hemoglobin's affinity for oxygen, known as the Bohr effect. As a result, oxygen is released to the cells and tissues to meet oxygen demands.

On a psychosomatic level, muscle tension may reflect deeper emotional or spiritual issues that need to be addressed and released. The exploration of the tension goes beyond scientific explanation. The scientific and spiritual reasons are not mutually exclusive and can both be valid.

What effect does smiling have on my experience besides going in full-heartedly?

The autonomic nervous system is complex and controls the involuntary physiological processes of the body. It operates all the time, and is involved in our decisions, what we pay attention to, our heartbeat, the pleasure and arousal that we experience. The autonomic nervous system control is divided into two branches: the sympathetic and parasympathetic components. Smiling stimulates the vagus nerve of the parasympathetic nervous system, expressed as an increased heart rate associated with facial expression recognition.[10]

The vagus nerve plays a role in cardiac and gastrointestinal function, in the neuroendocrine-immune system, and it regulates emotion. It also supports the production of white blood cells with new, enhanced DNA molecules that are programmed by the electrochemical signals of your thoughts.

Is this method safe for pregnant women?

Pregnant women should not be doing this breathwork so as to avoid the inadvertent adrenaline effect on the development of the fetus, which has its own primal relationship to DNA.

Can I do this with a partner or a loved one?

Yes. If you both are in an emotional space where you are open to exploring the power of this healing in your relationship, I would recommend lying side by side to do this breathwork together. If you would like to feel the energy that you create together, gently hold hands and pay attention to the sensations, such as heat, in your connected hands. There is a difference between your free hand and your shared hand. Be a witness to the evidence of the shared loving energy. Enjoy its beauty.

Why is holding my breath helpful?

There are a number of reasons why intermittent hypoxia is beneficial: it improves the immune system's ability to recognize threats within the body, enhances DNA memory, improves dopamine synthesis and distribution, increases serotonin, and stabilizes the nervous system. Box breathing is a form of intermittent hypoxia and a stress-reducing technique. It is even used by the U.S. Navy SEALs.

You are producing more red blood cells to transport oxygen efficiently, more mitochondria to create energy, advancing DNA material, reducing inflammation and cortisol (the stress hormone), and supporting the efficiency of your stem cells to regenerate your body.

How will this help my archetypal DNA?

Regardless of what your body and your mind make of the sensations, the point in this practice is to consciously synthesize new DNA with programmed thought using the wholeness of your conscious mind. Archetypes and consciousness transfer through thought and become accessible to your evolution in separate practices in your life.

If you use a smartphone, you typically have an option to program your voice or do a face scan to automate the ability to open programs using the technology of your sounds or light. It isn't any different from using your breath to create new molecules that will operate on your behalf, based on the programming you give them in meditation or breathwork. In fact, the subtle energy of breath and DNA create smartphones through the illusion of mind-drive technology. This breathwork overcomes fixed thoughts or interpretation of the former understanding of DNA. We are enhancing our own genetic technology through human and cosmic breath.

Dreamtime and the Ancestral Realm

When we dream, we're awakening the memory of our ancestors and their stories to remember our legacy. Our unconscious realm is more real than reality.

THE ASTRAL DIMENSION IS THE ANCESTRAL PLANE

Each person and animal possesses an astral body composed of its spirit's memory and connection to a cosmic source within its DNA. Because DNA contains evolving inter-dimensional transmission patterns, the astral plane is a consciousness sphere too. An astral projection is a type of intentional out-of-body experience as well as a lucid dream outside-in.

The astral body is a type of polarity that is difficult to convey or teach within spiritual contexts. The significance of this form of spiritual polarity is a felt experience of conscious meditation and channeling. We know from chemistry that polarity helps us interpret reactivity, the non-covalent bonding characteristics of an entity, and how it interacts with its surroundings. Similarly, the felt polarity of the astral body unveils something that is so vast and beyond our understanding of chaos theory, thermodynamics, or meditation. It is the field in which we experience an aspect of our human body being neither solely physical nor spiritual.

Humans exist in two realms. One accounts for all of our waking experiences, and the other one takes place in our unconscious, which includes subconscious, super-conscious, and trans-conscious realms. By trans-conscious, I mean straddling mental and physical realms. Both realms recognize visual images, metaphors, experience, and imagination. In this chapter, I refer to the following states of consciousness:

- The **conscious experience** is the observation of being through the physical body or waking state. It is characterized by awareness of light, sound, and breath.

- **Lucid dreaming** is the conscious observation of sound and light in a sleeping experience. It is, in a sense, knowing that we are dreaming within a dream. It is not a skill that most are born with but can cultivate. As a child, I had the same recurring dream of being separated from one of my parents. It came in different forms and settings, but the meaning and construction of the dream was very consistent to me. I woke up feeling scared, sad, and appreciative of their presence. The memory never faded, so I realized that the way I felt from the dream state was real. That is, it transcended whether I was separated from my parents in reality because in dreams I was. For being separate realities that dissolve into waking, dreams are no less real. Knowing that you are dreaming in a dream is living in sound and light.

- Only fragments of the **unconscious experience** are revealed to us by our consciousness.

- A **trance** is a relaxed, meditative state that allows light and sound to flow freely without attachment. This state is similar to the transition from wakefulness to sleep, where we shift from a physical to an energy body—a non-physical aspect of our being that experiences imagery, sound, vibrations, and movement while feeling detached.

- In a **telepathic exchange**, we anticipate tracking information from a distinct source, while accepting that we aren't separate from the source. Our inherent telepathic ability is its resonance to its auditory surroundings because we *are* a manifestation of sound.

- In my view, an **out-of-body experience** is a state in which our mind is able to differentiate identity, location, and perception as separate from our physical body. Ordinary sleep can be considered an out-of-body experience. The mind uses sound to visualize vibrations—an OBE traveler strings herself on sound as if rising into the air and floating in time and space.

- The **astral body** assimilates consciousness from daily life and recreates physical attentiveness, feeling, and spiritual activity within a subtle non-physical form. This phenomenological state enables a person the sense of seeing one's own body from an aerial position or a third-person perspective, a type of out-of-body experience. Although there is more anecdotal literature than empirical evidence related to understanding the astral body, the key element about an astral projection or OBE is that it is a form of dissociation and transpersonal experience.

In *Bottoming Out the Universe*, Grossinger, citing John Friedlander, writes, "The Astral body is the most polarized energy field in our universe, for every speck of the Astral generates a polarity."[1] That implies that our dream events, insofar as they are astrally generated, express polarities in our emotional, psychic, and libidinal life. In *Sleep and Dreams*, Rudolph Steiner gives a sense of the power of the astral condition when he says, "The astral body, however, even contains what we have *not* experienced. . . . Our astral body contains all of mathematics, and not only what is now known, but all the mathematics yet to be discovered."[2] It is no wonder that the mysterious nature of dreams seem vast.

My development with my astral body was neither comfortable nor a simple "deeply relaxing" protocol of altered consciousness guidance. Similar to Plato's direction with understanding Forms, it involved a rigorous acceptance of life education and experience. There was a pattern to my former illnesses and psychological challenges, which at its worst led to temporary homelessness and a suicide attempt when I was twenty-six years old. I succumbed to my pain, anticipated and loathed it day by day. I was living in my body because it was all that I knew at the time. My dad's death in 2016 awakened me. Initially, I thought that I was being punished, but that was another false belief about the Universe. What started as

a rebellious coping mechanism became my practice: Now I celebrate the frame that includes both suffering and joy in life. Not an artificial happiness or make-believe story, but a genuine realization whereby I could turn hardships into opportunities, platforms for evolution. If I didn't extricate myself from a marriage, apocalyptic church, and job, my trajectory would have been very different. In committing to share sound from my soul (and not just the scientific voice that says sound comes from the larynx, diaphragm, lungs, spine, and abdomen), the inspiration of this book was born. My angels spoke to my being, even if I didn't understand it at the time. From this focal transit, I started to develop a relationship to my astral body in dreams and meditation with sound.

Sound is a medium for the astral body by evoking emotions and sensations in the physical and spiritual realms. Through experimentation, I have discovered that various instruments can enhance this connection, particularly those producing elemental sounds that resonate with intuition and clairvoyance. Some of my instruments are flat, such as leaves, drumming, jiggling seashells and singing bowls, while others are airy such as strings of a harp or handpan, chimes, and tuning forks. If the listener is receptive and open sensually, the sounds can be felt down to a cellular subtle level, and with intention and practice, can translate into astral projection. But they are not automatic. It is easy to tune out their subtle elements if you hear them as mere "music" and don't siphon them singly "in."*

During my initiation into sound bathing, I wondered "How could such simple, primal sounds activate my being so deeply on multiple levels?" But the simple and primal are what activate everything—a truth we often fail to realize. The astral body is propelled by desire, while sound healing calms the body so that we can *feel* our desires and their emotions and direct them. In a sense, the astral body follows the direction of the heart song. That's why people go to concerts. With the

---

*My first music track "Nothingness" was composed to sonify the voice of the wind, gong, and chimes to engage with elements of the astral body. It is a hybrid type of hypnagogic rhythm, infused with mountain rain nostalgia and the mystical heartbeat of the planet co-produced by Christopher Galiffa.

heart's intelligence of cosmic and cellular love, the astral body can be directed by sound as by the voice of a thought.

THE STORIES OF OUR ANCESTORS

The energetic and spirit presence of ancestors is attested to by the totems people place and cherish. They think that it is memory, not presence, but it is subconsciously both. Millions of people in every country have photographs or mementos of their ancestors sitting somewhere in their house.

Storytelling conveys that a memory holds wisdom and a deep sense of meaning in everything that we do. A story has a life of its own, which becomes evident in our dreams and cultural mythologies. A myth is a condensed account of events across our many bodies and states of consciousness aggregated in the personification of a god, hero, or an archetype. Dreams and traditional mythology are gateways to access memories through archetypes.

The Mesopotamian "Descent of Inanna into the Underworld" is a story of sibling love and tragedy in which Ereshkigal kills and hangs her sister Inanna on a meat hook to rot. If you find this un-sisterly, know that it serves as an archetypal condensation of their shared suffering. Ereshkigal, tormented by physical and psychological pain, finds an unexpected ally in little flies who willingly listen to her affliction with empathy. Feeling seen and heard, she begins a journey of healing and asks what she can do for the flies in return; what can she do to show her appreciation and gratitude for the transformative experience she felt from the compassion and empathy that the flies showed when she was filled with remorse and shame? "If you are gods, I will bless you. If you are mortals, I will give you a gift. I will give you the water-gift, the river in its fullness."[3] The flies ask for Inanna's corpse in return and help resurrect her to life.

The killing of Inanna is the mirrored suffering of Ereshkigal. While rotting is commonly associated with death and decay, it is also a biological state of healing, birth, and new beginnings. I treat decay as surrogate life after death, one of the themes of this story. The flies share compassion for the hurting sister, perhaps even the spirit of "In-anna" saying "I see you" to her as a healing sound, a mantra and vibration. In the end, both sisters are brought to life. Although this myth has been told numerous times, its essences can be felt and experienced anew as if in a repeating dream.

DREAMSCAPES

When we sleep, our unconscious memories tell us how we truly feel about them or what they really mean to us. We go on an inner quest through a replica landscape. Our dreams are a vulnerable reality where we confront fluid energy patterns in their incompleteness. The words *hypnagogia* and *hypnopompia* derive from the Greek and mean, respectively, "to lead to sleep" and "to send away from sleep." In other words, hypnagogia is your pre-sleep threshold, and hypnopompia is your dissolving dream experience as you wake up. Both are semi-lucid with aspects of both waking awareness and dream consciousness, leading often to strange hybrid phenomena—visions and fantasies—the ability of the mind to recall in mind.

I believe that our dreams are, far from being enigmatic forms of intense, almost crystalline clarity. They reveal our mind's wild and unknown landscapes and thought forms, its fears and passions (Freud called them wishes in deeming dreams wish fulfillments) and capture an elusiveness that longs to be expressed. Dreams are like holographic highlights of our soul's news and updates.

Two-Way Communication in Dreams

In the movie *Inception,* Leonardo DiCaprio's character enters into other people's dreams to extract secrets from their subconscious minds. What was a fantasy science-fiction plot in 2010 has become reality in 2021. A research study involving four independent teams in France, Germany, the Netherlands, and the United States showed that two-way communication is possible in lucid dreaming. Scientists monitored the brain activity, eye movements, and facial-muscle contractions of thirty-six subjects with electroencephalogram helmets. Within fifty-seven sleeping sessions, six individuals were able to signal that they were lucid dreaming by moving their eyes and facial muscles in specific patterns such as a predesignated number of times to the left or right. Participants were also given simple math problems, like four minus two. The dreamers would signal the answer based on what they were taught before falling asleep, such as smiling or frowning or eye movements that matched Morse code. Across all subjects, nearly 18 percent of the dreamers responded correctly to 158 occasions during signal-verified lucid dreaming.[4]

Sacred Knowledge Shared in Dreams

Our dreams are channels through which sacred beings, or our ancestors, communicate wisdom to us. We are the receivers of their whispers of knowledge. An isopsephy reveals this truth: In the Greek language, *Ta oneipa* means "the dreams" and has the gematria value 537. The value is shared with two other words, the Latin *dektēs* and Hellenic *Iepa onta*, which mean "receiver" and "sacred beings," respectively. Through our dreams, we are the receivers of wisdom.

> Receiver = Sacred Beings = The Dreams
> **Hellenic: ΔΕΚΤΗΣ = 537**
> Latin: DEKTĒS
> English: Receiver
> **Hellenic: IEPA ONTA = 537**
> Latin: IERA ONTA
> English: Sacred Beings
> **Hellenic: TA ONEIPA = 537**
> Latin: TA ONEIRA
> English: The Dreams

Even scientific mysteries have been solved through dreams. Michael Faraday discovered benzene in 1825, but the structure of the organic compound C_6H_6 remained unknown. It wasn't until 1865 that the benzene ring structure appeared to chemist Friedrich August Kekulé in a snake dream:

> There I sat and wrote my Lehrbuch, but it did not proceed well, my mind was elsewhere. I turned the chair to the fireplace and fell half-asleep. Again the atoms gamboled before my eyes. Smaller groups this

time kept modestly in the background. My mind's eyes, trained by visions of a similar kind, now distinguished larger formations of various shapes. Long rows, in many ways more densely joined; everything in movement, winding and turning like snakes. And look, what was that? One snake grabbed its own tail, and mockingly the shape whirled before my eyes. As if struck by lightning, I awoke. This time again I spent the rest of the night working out the consequences.[5]

Kekulé's revelation isn't the only example of a classic scientific discovery that took place in a dream. Dmitry Mendeleev experienced a logical arrangement of all the chemical elements in a dream.[6] René Descartes determined the basis of the scientific method that came to him in dreams he had on November 10, 1619.[7] In an intense dream caused by tropical fever, Alfred Russel Wallace discovered the theory of evolution by natural selection.[8]

DREAM PRACTICE

I want to share how you can take an easy, step-by-step approach and make it your own dreaming practice, lucid and positive.

Create a Goal: As in dreams themselves, your goals should include your desires, will, and curiosity. What do you wish to experience in your dream? What do you want to visit or see? Are you trying to connect with a loved one in another dimension? Do you want to enhance your creativity and imagination? Be clear in your intentions. "I want to meet my mom and be open to the wisdom she has to share with me so that I can be motivated in the next step of my life." The way you ask your dreams to help you receive this knowledge is an agreement within your own mind. It isn't any different from talking with your friends, animals, plants, water, or DNA—intention is the driver. When we interact with an unknown, chaotic form of energy in dream space, we consciously lead with a clear intention to remain productive. Our dreamtime experiences are shaped by the way we treat our dreaming mind.

Create Expectations: Allow yourself to feel joy, excitement, anticipation. This is a critical part of your body's intelligence in priming for lucid dreaming. Even if you don't feel it at first, tell your mind, "I'm ready!" Your body will begin to feel it first. The way you treat your body can also affect your lucid experiences. Drink water and avoid eating late at night. In fact, you can prime your body for lucid experiences by going to bed hungry. Become a hungry prowling bear rather than a soggy hibernating one. A meditative practice before bed is an aid.

Sound healing can also be a way to ease into lucidity. When I teach sound healing classes in the evening, I typically have a lucid dream the same night. Some of my participants self-report having a lucid dream after a sound session as well. Preparing your mind with images, sounds, and smells of nature or movements before bed amplifies body awareness. Watching a documentary about nature or taking a walk or stretching in the evenings sometimes helps us feel lighter and able to move around with ease in dreams. We are able to transfer the cinema images into our dream body and they become dream movements in lucid dreams.

Containment: When you realize you are lucid in your dream, the next step is containment: attuning the awareness in your unconsciousness to developing range. This can be tricky. It's easy to get too excited and wake yourself up. Practice staying calm in your daily regular life as a dry run for staying in your lucid dream life. Recognizing your excitement as "new" is all part of the process. The key is learning how to overcome strong emotions. In your waking experiences, practice taking long deep breaths or doing an activity that calms your emotions. This could be time in nature, meditating, doing conscious breathwork, or any mode that helps you achieve authentic tranquility.

In the process of developing your sense of calm, actively interact your surroundings. By engaging with your environment, you express curiosity about its lucid nature without straining your focus on a single technique. Avoid staring or fixating on a particular point. You can look at your hands or feet and keep moving your eyes around your surroundings.

When you get to a point where you are aware that you're dreaming in a dream, move around and interact. You can begin by simply listening. You can try to walk, jump, or touch the actual aesthetics or mechanism of the dream. Engaging with the dream is the way we cultivate self-awareness. Learn to affirm your lucidity with simple statements such as, "This is a dream." When you understand that it's a dream, the resolution of its imagery becomes more responsive, even telekinetic. The best way to capture its stimulation is by repeating whatever you are doing in your dream to keep you grounded and focused. For instance, if you're in a house with four windows, and the wall color is orange, say almost pedantically, "I'm in a house with four windows, and the wall color is orange." Speak the details so that they double or reverberate and you can stay focused on your dream.

INVOKING THE ANCESTRAL REALM THROUGH DREAMS

∽o∾

My awareness of ancestral dreaming began after my dad died. For a year, the childhood nightmare of being separated from my father activated. I spent a lot of time in phases of sleep not just to recover from grief, but because he appeared in each dream. I desperately tried to hold on to him and pull him back into this world. With time, my perception of dreams shifted. I realized that they could not be resolved on their own, but were a necessary element in my evolution and healing process.

Tune into your dreams to awaken your memories of the ancestral realm. Let the voice of your memory be the voice of your ancestor. It is evident that we are capable of experiencing two-way communication in our dreams by and through the ambient nature of our surroundings. I believe our astral body is capable of connecting with the ancestral realm. In my dreams, my dad and grandmother reveal future events, such as pregnancies, unfoldings, and personal experiences. This is a gift you can cultivate as well.

When I dream, I depart from this human body while my astral body activates an unconscious memory. If I am aware of it in my dream, I will bring it back in some form to my waking experiences. This discovery led to my ongoing journeys as a disciplined dream traveler. I learned to shift my mindset in nightmares, welcoming them as opportunities for growth. In the lucid state, I honored the dream channel as a mode of communication and a guide in "letting go" in an unconscious, dark realm. When I treated the dream as communication, the perceived nightmare dissolved into pure information. At the same time, I was developing a relationship to my astral body in my waking experiences using sound.

Ancestral dreams can be seen as a type of out-of-body experience (OBE). Before you explore what an out-of-body experience is, ask yourself: Are we separate from our dreams and sound? Dream characters within our lucid dreaming experiences will not interact with us if we believe that we are separate.

When I generate a soundscape in a classroom setting or play an instrument, it is its own mild out-of-body experience. I do not see bodies in or beyond these instruments. I see musical gods that demand being played. The sound that is created is divinely sacred and part of the audio temple of those who seek its healing. That's why people go to concerts of every sort, from Bach to electronic music, country music, jazz, reggae, rap: to be healed by sound frequencies, to take a lucid sound-meaning bath. It is a type of snake charming, except the snake is the frequency of the instrument, the sound is the healer. When I walk around with my harp during a sound session, each of the thirty-six strings' vibrations resonates within the anatomy of each individual around me. I repeat the method with each instrument. At the end of the session, we sit or lie in silence.

People often leave the classroom in silence or whisper "thank you" or tiptoe on their way out as if they were in a holy temple. It is an aspatial and atemporal temple in a celestial sphere. It is where the astral and physical planes meet.

Sometimes it can be discouraging when the intention is to invoke the ancestral realm in our dreams, but the dreamscape is quite different from what we planned. Remember that when we think we seek our ancestors, they are seeking us too. Here is a practice to prime your being and initiate an ancestral dream beginning with the waking realm.

Engaging with Your Astral Body

A simulation can help prime your technique in doing an astral projection. Meditation does not necessarily mean your astral body is literally performing what is described. Think of it as a bike with training wheels. If you can visualize this experience, you can start to experiment with your individually curated astral meditations.

Close your eyes and allow yourself to relax in the normal rise and fall of your breath. Pay attention to the sound and passage of your breath. Allow yourself to sink into the moment by listening to the beat of your heart, the flow of your breath from your nose, and the soft whisperings in your lungs. When you start to dial into the sensitive sounds of your breath, you may begin to notice the other subtle sounds of your body, including the movement of your hair strands, the sounds of your jawline opening and closing, the saliva in your mouth, your fingers gliding along the skin of your legs or arms. Yes! Each of these soft yet distinguishable sounds of the body help dial into the astral body.

Imagine the emergence of your unconscious' consciousness. It won't happen immediately, so be patient. If it doesn't come at once, ask it to appear. Even if it doesn't have a shape or form, you can feel it, and this is a starting point. *Ask.* With your eyes closed, visualize your astral-ness emerging from your body above you. Let your astral body look down on your physical body. What does it see? What features of your body does it dial in? Watch your astral body float around your body and then explore the room. What does it see in the room? Your astral body sees the space of the room. If you don't see it, imagine it. Write its choreography or script.

Continuing in your normal breath, see your astral body leaving the room, out of your place of dwelling. Let it transition outdoors. Space-time doesn't curtail your astral body's capabilities for movement. See your astral body travel long distances within moments. Its awareness shifts from your residence to signs, roads, people walking on the street, tree-top canopies, and seeing the horizon of the sun—that is conscious unconscious movement. Your astral body sees the peak of a mountain on the horizon and, within an instant, finds itself there.

Now the astral body is in the mountain range. Have a look from this aerial viewpoint through your astral body and its cosmic source. See a retreat-like dwelling resonating with rainbow frequencies not far from where you are. Watch your astral body travel there. Have a look at a heavily wooded lakefront building and landscaping with wild grass, gardens, crystal-clear water, sacred sculptures, and inter-dimensional energy movement. See the sleeping porches and sunken courtyards as you navigate the campus, allowing intrinsic light to lead your way.

Allow yourself to enter the building and go to the center of it. There is a spa pool where you are invited to sit and receive energy frequencies in a vibrational bath. This frequency will give you nourishment and relaxation. It is similar to sitting in a warm pool, in this case a pool soothing your astral body. See yourself immersed in it. As you melt into this field, notice the sound of your breath calming, more so as the resolution of your experience becomes better defined. You may start to see golden or blue energies move through your field. Surrender to them. They are where you need to be.

After a few more moments, you'll start to see water draining from the pool. It wants to pull toxins from your physical body. Let it. You may feel newness in your conscious body from astral projection because your thoughts are eliminating toxins through active visualization and a kind of psychic neurolinguistic reprogramming. Take your time to move your astral body gently. Drift, don't herd it. Go back outside the imaginal retreat building and breathe. Shake off any energy that says this is outside of your comfort zone. This is new.

It is time to return. Imagine your astral body traveling back in the same way it arrived. Pay attention to any details that you didn't see before, any resolutions that make themselves more clear. That establishes their out-of-body reality. See your astral body return to your residence in the physical realm, greet your physical body, and reconcile back.

Open your eyes gently. Take your time in moving physically again. Slowly blink your eyes. Moderate your inhale and exhale. Wet your lips. Drink some water. Take your time moving so that you complete an astral projection but also heal yourself. Thank your physical and astral body.

How did your body and mind feel before you began?
How did your body and mind feel after the journey?
What differences do you notice in your post-conscious review?
What memories were activated?

The Step-by-Step Ancestral Dream Ritual

For this ritual to explore the ancestral realm in your dreams, use *Artemis vulgaris*, or mugwort, to assist with lucid dreaming. Be aware that some individuals have reported experiencing darker dreams when using this herb—dreams that reveal hidden insights and core wounds. Please research this herb or any substance that you plan to consume and, when possible, grow them yourself.

You will need:

A journal
2 tsp of dried mugwort/teabag
Hot water

Before making the tea, set a strong enough intention. Journal your intention prior to making the tea and drinking it. Invite the ancestor from either mother or father's side that wants to be with you now. You can even journal this as a question, "Who is the ancestor wanting to be here with me now?" Invite them in with respect into your space. Use your intuition, feel their spirit. Feel them as a younger version of themselves. Write down these questions into your journal:

What was your life like?
What were the challenges and struggles in your life?
What are your dreams and aspirations?
What needs to be seen and heard in your life?

Take a deep inhale followed by a long exhale, soaking in the moment. Call on your divine guides to guide you in this dream ritual. Let them be the matrix that supports the dream. Express gratitude and desire for grounding.

When you lie down to close your eyes, establish a comfortable rhythmic breathing pattern. It is almost like *savasana* in yoga. Relax your jaw. Let go of any thoughts that you have about your day. Let yourself drift off and know you're safe.

After waking up from your night's sleep, see if you can answer the questions in your journal. Document your dream and record anything you remember, including location, people, items, and actions. Do not derogate anything that negatively came up for you. Hold the dream with grateful regard and friendliness. You are on your ancestral dream journey and learning more about the memories of who you are. There are no bad adventures; that's the value of lucid dreaming. For that matter, it is the value of lucid living.

Reflect and journal the elements of the dream that inspire gratitude. Even if you didn't have your desired ancestral dream, be patient. Your ancestors will reveal the answers to you in a separate waking or sleeping experience. The conclusion will come with time and will also continue to evolve.

When I enter a session where the intention is to help a person heal, I open my heart with complete vulnerability. In a sense, I do experience a particular type of fear. This fear is meaningful and good-natured because it comes from a place of respecting the profound life and evolution that wants to radiate through this timely process.

Each individual's healing journey is authentic, which makes my role both exciting and demanding. I must maintain a clear energetic channel to create a memorable experience. I must be gentle but structured enough to guide the frequency of the other person the same way a bird finds its own lift using the wind beneath its wings. Trust is a crucial element. A person does not only share one's own sensitive experiences, but primarily, activates a unique and authentic evolution cycle. This involves a simultaneous death and rebirth, the mourning of a loss yet an ecstatic joy in new life. At a baseline, it is psychically complex and challenging when both are felt at the same time. This is where my role is akin to a spiritual death and birth doula. I help people feel empowered and enchanted with their individualized consciousness and experience a wakening zest for life.

12

Composting Grief

Grief reminds us that, in the divine alchemy of
transformation, things sometimes appear to go backwards
as they move forward.

The rewilding of humanity and our genetic transformation is contingent on the way we handle grief. Grief is felt as a loss of species, ecosystems, and meaningful landscapes due to an environmental lesion. It constantly shifts and changes, varying from person to person at different stages of life. Grief is a separate political system, indifferent to Left or Right ideologies and solely focused on consciousness.

All levels of consciousness are affected by grief, whether it is emotional or mental, ancestral or generational, ecological or cellular. A grief-related thought can alter our sound, vibration, and even the functioning of our DNA. While grief cannot be contained or fixed, we can learn to walk alongside it with strength, calmness, and order. I approach healing with a variety of techniques to help you enrich your soul's relationship with your DNA, Gaia, and sound: a trifecta. In this chapter, we'll explore a blend of scientific and holistic principles to alchemize a divine approach to honoring and transforming grief in a new way.

THE GRIEF EPIDEMIC

Grief is like a virus. As a scientist, I studied the sensitivity of cells to viral infections. Viral infections typically destroy the host cells as they are overtaken by the virus's progeny. It does this by integrating its DNA into the host-cell genome, replicating itself using the host cell's own machinery, and spreading further infection. Viruses aren't the only organisms to evolve numerous strategies to create cellular genetic damage; grief leads to a plethora of consequences for genome integrity and should be considered a disease mechanism.

Although how grief integrates into our DNA is obscure, it is marked by stress and trauma within DNA. Similar to a viral infection, grief can have multiple consequences for host cells, leading to cell death and tumorigenesis, and affecting our genetic evolutionary participation and future generations. However, if we decide to shift our energetic trajectory with grief toward healing, it can contribute to shaping our genome.

Before we go into managing grief, I want to emphasize the importance of recognizing stress. Stress smothers memories, irrationalizes thought, catalyzes abnormal growth, and reduces life's vibrancy in every sensation. "Heart disease, stroke, and cancer are the leading causes of death in the world" is a filtered version of saying that the imbalance of stress has fatal consequences on the integumentary, skeletal, muscular, nervous, endocrine, lymphatic, respiratory, digestive, urinary, and reproductive systems. But how does grief and stress affect cells, leading to all of these disorders? It starts in a specific type of DNA called *mitochondrial DNA*, the specific type of DNA found within our cells' energy-producing structures.

MITOCHONDRIAL DNA DEFECTS: A BIOLOGICAL PATHWAY OF GRIEF

Mitochondria are organelles within each cell known as the cell's powerhouse, providing energy for biochemical processes and converting energy to drive reactions and oxidative transfer. Believed to have once been an independent organism that was evolutionarily and opportunistically engulfed into our cellular machinery, it is unique as it contains its own DNA, separate from the rest of the cell's genome, known as the mitochondrial DNA (mtDNA). Inherited from our mothers, mtDNA is enriched with structural and electric mechanisms from its bacterial origin.

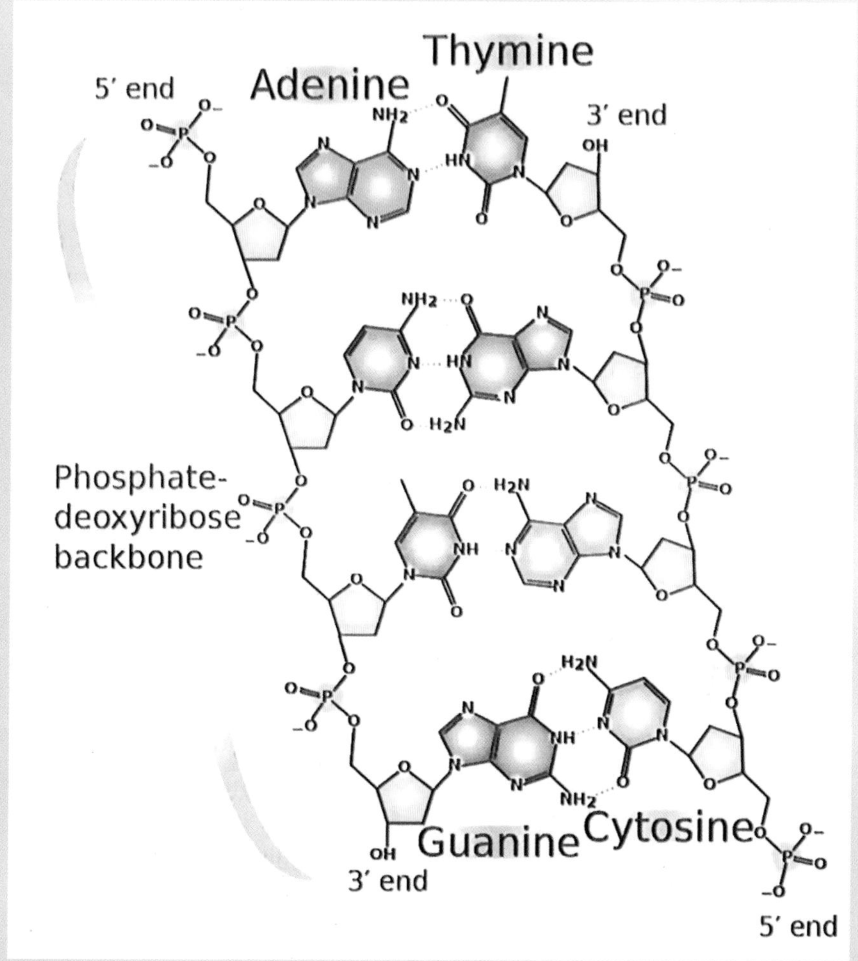

Plate 1. Chemical structure of DNA, hydrogen bonds shown as dotted lines. Each end of the double helix has an exposed 5-prime phosphate on one strand and an exposed 3-prime hydroxyl group (-OH) on the other strand.

Illustration by Madeleine Price Ball

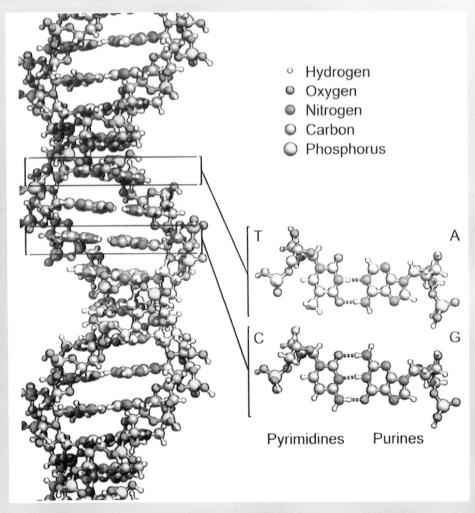

Plate 2. The detailed composition of DNA reveals its structure
containing four distinct bases: adenine, cytosine, guanine, and
thymine, and their purine and pyrimidine forms.

Illustration by Zephyris, 2011

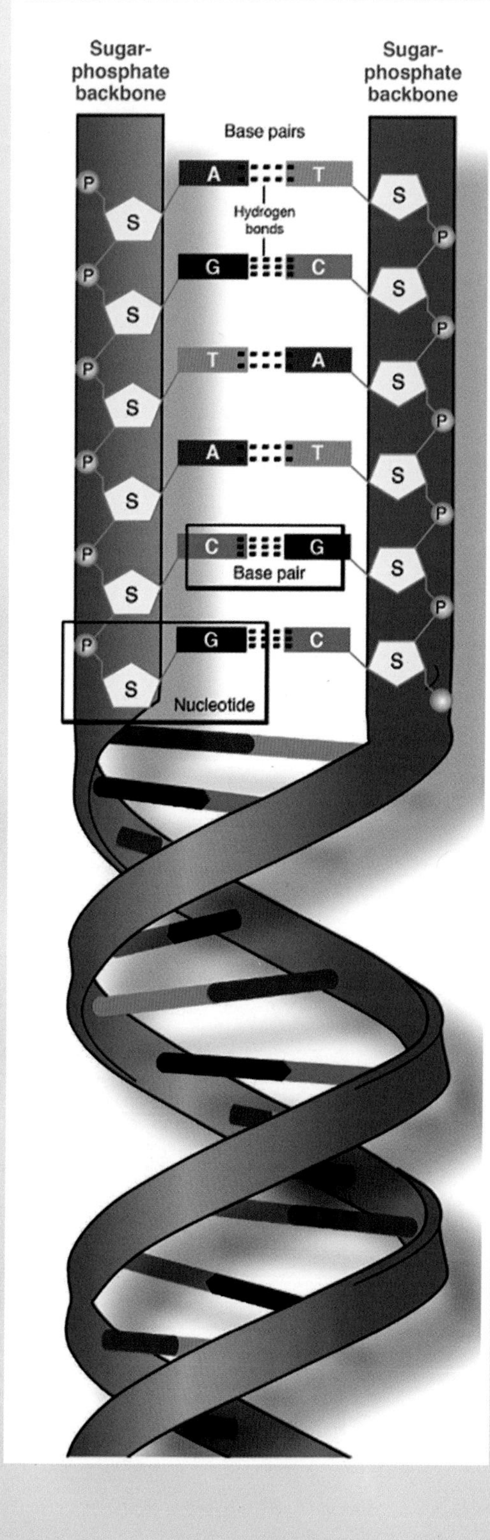

Plate 3. The DNA structure diagram depicts the phosphate backbone and its relationship to the nucleotide base pairs.

Courtesy of the National Human Genome Research Institute, genome.gov.

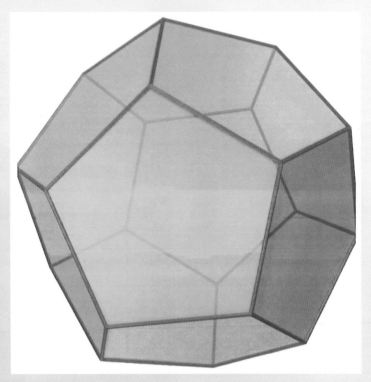

Plate 4. A Dodecahedron; a regular polyhedron.
Illustration by Cyp/Poly.pov

Plate 5. Professor Robert Langridge's dodecahedral DNA.

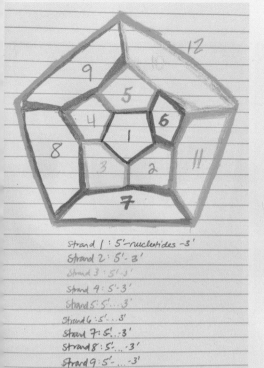

Plate 6. Our 12-stranded DNA.

Strand 1 : 5'-nucleotides -3'
Strand 2 : 5'- 3'
Strand 3 : 5'-3'
Strand 4 : 5'-3'
Strand 5 : 5'...3'
Strand 6 : 5'-...3'
Strand 7 : 5'...-3'
Strand 8 : 5'...-3'
Strand 9 : 5'-...-3'
Strand 10 : 5'-...-3'
Strand 11 : 5'-...-3'
Strand 12 : 5'...-3'

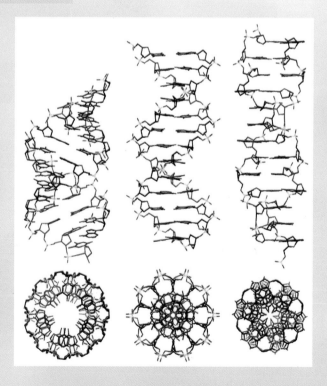

Plate 7. From left to right: A-DNA, traditional B-DNA, and Z-DNA. The middle form, B-DNA, illustrates the traditional double helix. Each DNA structure exhibits a mandala-like appearance inherent within its molecular arrangements.

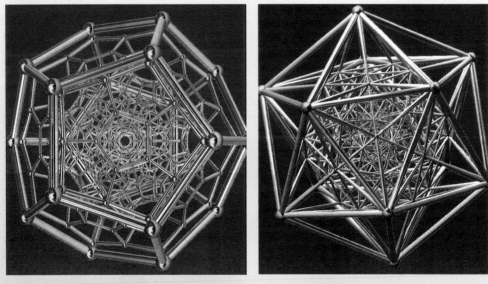

Plate 8. Schlegel diagram, by symbol. Left: 120-cell, {5,3,3};
right: 600-cell, {3,3,5}.

Courtesy of Robert Webb's Stella software

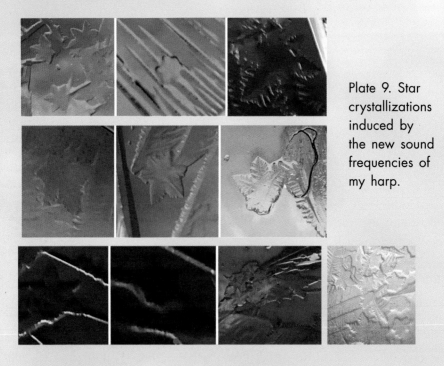

Plate 9. Star
crystallizations
induced by
the new sound
frequencies of
my harp.

Plate 10. Water reflecting stress.

Plate 11. Water lotus

Plate 12. Water speaks evolution.

Plate 13. The mermaid in the water seems to be reaching its hand out of the water.

Plate 14. Being beyond double helix: I took this photo at the Navajo Nation Parks and Recreation. The image invites the viewer to contemplate the wonders of nature.

Despite having adapted over thousands of years to become an integral part of human physiology, mitochondria have an Achilles heel in that they are vulnerable to oxidative stress. When there is an imbalance of antioxidants to free radicals, oxygen-based radicals are produced. The bane of aerobic species, these free radicals damage DNA.

Grief itself contributes to oxidative stress in the body, and the mitochondrion is a prime *target* of grief as well as a moderator of grief pathophysiology. MtDNA is densely packed with exons and lacks introns, the noncoding regions described in chapter 6, making it more vulnerable to alteration and damage in comparison with the nuclear genome when it is subdued by grief. Mitochondrial dysfunction can impact various organs and tissues, such as muscles, kidneys, the liver, and the brain. The severity can translate generational trauma that manifests in different emotional and physical expressions such as obesity, autism, asthma, and attention deficit disorder.

There *is* a healthy way of transmuting stress into growth, but it requires us to recognize the way we look at grief. Grief can be a powerful ally in a new system of consciousness.

GRIEF
Ally and Divine Force

Works of ancient Greek literature explore grief as a divine force. In Homer's *Iliad*, one universal right is the memorial. In the Trojan War, silver coins were placed at burial sites of fallen soldiers as tokens to remember the lives that were sacrificed in service of a shared civic mission. At the end of the story, the physical tokens were replaced by mourning and shaped stories: a combination of totemism and ceremonialism. The narrative is repeated in Sophocles' "Antigone" and Euripides's "The Suppliants," which encapsulate the perspective that grief is a divinely sanctioned right. As a metaphorical cicerone, memories of the past continue to accompany our way through life and lead as a guide to the future. There is still a delicate balance in the way we treat our grief and memories of our ancestors and creation.

Grief doesn't enslave a person, it follows the person's intentions.

One doesn't need to be scientifically trained in ionic interactions to learn how to transmute electrophysiology. The key is in recognizing how profoundly our thoughts can shift our genetic state. We have an aptitude to evolve without waiting hundreds or thousands of years. I believe the human body is capable of communicating through signals of sound, hormones, and photons to itself and its surroundings, including underground ecology—the Earth.

A GRIEF ECOSYSTEM
Soil and Vocal Soundscapes

Nature's medicine lies in the ground. The soil is the dust, blood, flesh, and bones of our ancestors as much as it is a part of our own being and waking experiences. It is so present in molecules of air and food that we miss it entirely. The soil connects all of the pieces together, in a sense, the way grief does. A person's progression and relationship with love, truth, and goodness unfolds over a lifetime. In contemplating the source from which the rivers of this all-reaching evolution springs, it may be possible to examine grief by which the complexity of perspective, interconnectedness, and meaning continues to refine and fashion our intimate expression and embodiment of divine love. Grief is the intimate fabric of connection with the natural world. Humans belong to the land just like any animal, plant, mineral, water, and air particle, and each being experiences the emotional, spiritual, mental, and physical fertility of the ground.

The concept of "composting grief" resonates as a way of transforming energy into a renewed sense of purpose and growth. I think of this as transmutation or snake medicine: sound, meditation, selfless service, praying, nature, movement. I propose a scientific, physiological, and ethereal level of meditation for the symbiotic relationship between human grief that supports its own underground microbial life—its microbiome.

∽o∾

A book, Grounded: A Fierce, Feminine Guide to Connecting with the Soil and Healing from the Ground Up, *was sent to me as a surprise gift by my friend, Maggie Kennedy, whom I hadn't been in contact with in over a year. I received a message from her saying she sent me a book that she felt would resonate with me. Not only did it resonate with me, but it turned out that Erin McMorrow's book on sacred interconnectedness activated my own path of regenerative healing through writing. So, I became a writing student under Dr. McMorrow's guidance.*

"Composting is a form of giving back to Mother Nature and the cycles of nourishment," Dr. McMorrow says, "and it's also a reminder that the things we think we can't use can actually be returned to the earth with care and used for something else (the material version of releasing any energy from our bodies that doesn't serve us)."[1] Human and soil microbiome share an inherent quality: both are in a constant state of repurposing energy to regenerate life: "grounding" and "earthing." Studies have yet to explore the proliferating impact the human body has directly on the soil through meditation because it's immeasurable. One can only trust, believe, and commit to doing it to experience the results. By prayer, mantra, or meditation, human beings can create a meaningful relationship with soil.

Similar to composting, I believe a human being is able to cognitively and physiologically transfer ionic energy containing grief information into the Earth through conscious meditation and grounding techniques. What has been referred to as "grounding," or simply placing your bare feet on the Earth's surface, reduces cortisol levels, inflammation, and calms the nervous system through the absorption of the ground's negative ions. The soil contains various bacteria that absorb and masterfully manipulate electrons and ions in their environment. In meditation, we intentionally release feelings of hurt, worry, and fear, guiding our thoughts to calmness. Breathwork further enhances this process by encouraging awareness and presence within each cell, helping to alleviate anxiety and welcoming openness to new possibilities. Using the same vibration-driven principles present in grounding, meditation, and breathwork, a person reconnects with their inner self and facilitate an energetic exchange of grief with the soil, fostering personal healing and evolution.

The soil is a network of life-giving microorganisms—similar to the mitochondrial membrane potential—that support the vital bioenergetics of the underground mystery. Actinomycetes, fungi, bacteria, nematodes, earthworms, whiteworms, millipedes, flies, land snails, and beetle mites are first-level consumers of compost.[2] This layer supports the rest of the underground microbiome—protozoa, rotifera, soil flatworms, predatory mites, ants, rove beetles, feather-winged beetles, ground beetles, mold mites, springtails, pseudoscorpions, centipedes, and other undiscovered species.[3] All of these organisms support the existence of higher tropic life forms. They date back nearly four billion years and will probably survive any future extinction event.

This diverse soil microbiome mirrors our body's repair system. Whenever a cell exhibits genotoxic effects such as mitochondrial oxidative stress, DNA fragmentation, alteration of gene expression, or genetic mutations, T cells are recruited to mount an immune response. T cells work together with glial and neuron cells to regulate the damaged cells' microenvironment: glial cells are insulation and nutrition; neurons carry messages and meanings. T cells can be likened to soil microbes that repair soil health through nitrogen fixation and suppression of pathogens. Grief is cached inherently within the memory of T cells and turned into a passage of intergenerational information.

As we've seen, bacteria are an important component of the soil microbiome. Bacteria aren't solitary organisms. They exist on everything and are deeply embedded in the human DNA construct. They use a cohabiting language of electrical signaling to other bacterial forms for their survival. This mode of communication between humans and bacteria isn't fundamentally different from the language between humans and water, or humans to humans. It is as real as channeling, meditating, or talking with a loved one.

In 2020, one of the first studies examining the impact of meditation on gut bacteria, which are crucial to overall health, compared bacterial responses between study participants who meditated and those who did not. The results showed three types of bacteria—*Bifidobacterium, Roseburia,* and *Subdoligranulum*—responded to the brain's influence and led to a series of changes in the intestinal flora that improved the body's holistic regulation. These changes included enhanced cellular motility and stronger immunity.

Meditation activates the gut-brain axis (GBA), an increasingly discussed topic in the last decade. The GBA is a bidirectional communication network between the central nervous system, autonomic nervous system, and the hypothalamic-pituitary axis (HPA). The gut-brain axis integrates gut function with emotional and cognitive centers in the brain. A 2021 study on meditation and gut microbiota proposed that meditation can reinforce healthy gut-brain axis* crosstalk through the central nervous system (CNS) and ANS pathways.[4]

While conventional science emphasizes the role of the gut microbiome influencing our emotional well-being, my inquiry on the axis between meditation, emotions, and soil ecology is yet to be systematically explored: The study on the effect of meditation on intestinal flora hints at a possible connection between emotional well-being, as fostered through practices like meditation, and the broader ecological systems that surround us.

The link between meditation and the underground microbiome—composting grief—draws upon the relationships among cyclic meditation, the microbiome, the gut-brain-axis, esoteric teachings, and various healing practices. We've previously seen the significance of sound and vibration in healing, and composting grief is no exception. A 2022 study by Maeder and colleagues used acoustic "soundscapes" to predict temporal and spatial dynamics† of organisms in the soil.[6] Although soundscapes are more commonly researched above the ground, the results of the study suggest that the acoustic diversity can be calculated from different soil communities. The evident differences between soil soundscapes can provide insights into the soil's composition (such as taxa richness) and help evaluate the relationships

---

*The intersection of our gut, nervous system, and microbiome is creating a new paradigm shift in medicine. Numerous studies show stress alters neurotransmitter release and affects the GBA profile.[5]

†Spatial dynamics is the study of the relationship between an entity and its surrounding environment, and as a result, temporal dynamics is the observation of the inherent complex changes to the entity's behavior over time.

between living soil microorganisms with greater precision. In a similar manner, we apply eco-acoustic technology and statistical methos to observe the impact of grief meditation and sound as a noninvasive approach to monitoring our human "soil ecology" in the composting grief process. Grief reminds us that, in the divine alchemy of transformation, things sometimes appear to go backward as they move forward. Through our shared experiences of grief, both we as individuals and the planet can evolve together.

In the next sections, we will go over two levels of meditation to transmute grief through soil and sound. Allow yourself to sink into what feels natural for you. Healing is an undefined, nonlinear process.

Composting Grief Meditation

Start by putting both feet on the ground, or visualizing the feet in the soil. You could close your eyes and see roots from your tailbone lowering into the ground, physically connecting with the land. Think of gratitude for the soil and its underworld microorganisms.

Thank you, Mother Earth, for sustaining our bodies and spirit with the sacred spirits of water that I return back to the land, rivers, and streams with gratitude.

Call on your divine guides, ancestors, and angels.

Reflect on the grief that you would like to compost: fear, loss, shame, jealousy, anger/rage, anxiety, depression, hopelessness, confusion, discomfort, chaos, hate, and discord—stressors that can be associated with grief and trauma. Form them into a shape. Take an inhale, and with the exhale, move the shape in a downstream pattern through your perineum, legs, and feet into the ground, like tree roots. Repeat with each inhale.

This motion creates a downward spiral, shifting laterally to cycle upward. The "down" cycle represents the grief that will be energetically decomposed by the underground microbiome to give life; the "up" cycle is the oxygenated life that is created by the microorganisms and absorption of the ionic charge from the ground. This circular scheme is stimulating a cellular connection between the human body and the underground microbiome. Its biochemical symbiosis will enrich the acidity, moisture, and oxygen to allow decomposition slowly over time with safe byproduct levels of carbon dioxide. It is earthing for you, composting grief for the fertility of the soil and the garden of your life.

You can stay in this meditation for as long as you like. Gratitude is the portal through which your emotions continually filter through as they penetrate the ground through the biochemical channel of your chakras. When you are ready to end the meditation, give thanks to the land and your guides.

Placing your hand on the ground is a way of relaying the ground's sensitivity to the brain. The ground will share information to you in multiple ways, whether through its vegetation, animal attraction, or the vibration that the ground connects with you on a cognitive level. The more you practice it, the stronger the body's efficiency with the ground becomes. Remember, this practice won't fix grief; its scope of healing is to transmute energy into restoration and prolificacy.

Grief Medicine Using Elemental Sounds and the Voice

The crashing and swelling of water sounds, such as the sea or ocean waves, is naturally calming to the brain, for it contains the delta and theta rhythms of its oceanic origin. The brain's activity behaves in wave-forms called neural waves or brain waves as discussed in chapter 9. Mimicking the sound of water initiates brain processes that regulate the levels of cortisol, serotonin, melatonin, and epinephrine in the body, particularly the heart where love and grief are felt. The exchange of sound in the brain's neural waves isn't different from plasma or seawater carrying oxygen, nutrients, and minerals from one cell to another.

The voice is a channel for communication and heart expression. We can use our voice to make water sounds that relieve and express the desire of the heart. The next exercise activates the sensory cells of the heart by the breath, voice, and tailored sounds of water from the sea. It is intended to wash out grief-related toxins from the heart.

Box breathing is used in this practice because its tempo is slower and easier to incorporate when adding in vocal expression.

Box breathing summary:

Step 1. Breathe in counting to a count of four slowly, filling your lungs with air. Hold your breath for four seconds. Slowly exhale through your mouth for four seconds and then hold your breath for another four seconds. Repeat for a couple of rounds, without holding your mind to any particular timing. Let the flow of the breath determine it. The goal is to attune your breath to behave like a wave of the sea.

Step 2. Visualize the sea with its perpetual cycle of waves. Each wave is unique and different from its former. Imagine the different kinds of sounds the waves make. They can come in forms of whispers, hissing, clicking, thuds, tapping, croaks, grunts, moans.

Step 3. After your second or third round of box breathing, slowly emit a sound that mimics the ocean. The easiest way to transition into this step is by starting with a hissing sound or make the sound of S and see where it takes you. Feel that you freely change your mouth shape. Extend the jawline, clench your teeth, grin widely, form a circle with your lips. Even if it feels exaggerated, do it to activate the memory in your anatomical memory. The cells remember.

This may feel unusual and awkward because people do not intentionally make such shapes in their ordinary conversations. Allow yourself to feel that you can be free and let the sound go. Take a huge sigh and practice letting go. It may help to incorporate your voice into it with no judgment. Affirm that your voice sounds right just the way it is. Make a different sea sound on each exhale.

Step 4. After you feel that you've exhausted your imagination of sea sounds, relax your jawline, mouth, and lips. Take a deep breath, let your belly relax and fill with air. On the next exhale, naturally decompress your lungs while releasing the air from your belly through your diaphragm and emit the sound that comes from your heart in your current state. Don't over-think it or connect to *how* it sounds, but let the heart express itself through this warmed up channel. Repeat for a few cycles, maybe even slowing down the pace to connect with the sound of your voice to feel the frequency in your heart.

With intuitive nutrition, elemental sounds, and effective fertilizing techniques, our understanding of DNA and its potential for adaptation continues to evolve. This deeper understanding enables us to cultivate a symbiotic relationship with grief, transforming our perception of its role in our lives. As we learn more about bacterial intelligence and its impact on consciousness, evolution, and cognition, we discover new ways to relate with our bodies, nurture the land, and pursue a path of mindful individuation.

Animal Medicine and Holism in the New Age

An animal, a holy being, isn't any different from our intrinsic connection to angels and our breath. Anima itself means "having breath" and "soul."

A Messenger of Illusion

I found myself sitting in front of my dad's memorial, watering his flowers and grassy area with a bottle of water infused with my mantras and a parade of my tears. In the brightest day with no cloud in sight, the sun shared its invincible joy in many rays to all of life. Except mine. His engraved portrait resonated the same warmth and comfort, as it always does on the black glazed marble stone. While my life shows the growth and evolution in the way I wanted it—personally and professionally—my deepest desire was to share my joy and gratitude with him. He led my family out of Kyiv when Vladimir Putin started to come to power; he saw my future though I was only twelve months old. He was like Moses who got his people across the desert but didn't live to see the Promised Land.

As I wiped the tears from my face, my eyes settled on the Ironman medal that I brought back from Australia. I'd secured it on his porcelain vase attached to his memorial a few years earlier. A dragonfly sat on it, waiting to be discovered. Unbothered, it sat and looked back at the picture of my dad, with me. What did it see? Certainly not him as I saw him, but something, perhaps an aspect of the aura I was projecting. I took a scan around the other memorials, gardens, and grassy areas to see what else was in sight. The flying zooid with multifaceted compound eyes was a loose actor. It came to share its purpose and to comfort me, as it witnessed a woman's complicated heartache over her father's absence. Is "witnessed" the word? I wonder. It did something less but also far more. It may have wondered if I realized the illusion of death, causing great restriction and limitations between the mental and spiritual realms in my own actions and ideas. At the moment that I grasped this concept, it flew away.

THE WORD MEDICINE AND EVOLVING WITH THE ANCIENT ANIMAL LANGUAGE

Medicine is a term that can make a person grimace nowadays, either by its socioeconomic framing or in the way it values or devalues self-esteem. It's a self-proclaimed topic in every media outlet. Most of the world is familiar with what *medicine* means by its Westernized definition. I am using the word *medicine* in its traditional meaning, "to connect" or "a way of life." *Medicus* and *medicina* in Latin means "physician" and "the art of healing," respectively. In Sioux, Lakota, Cheyenne, Aztec, and Mayan teachings along with other indigenous traditions, medicine people were considered the most spiritual and divine. They healed by invoking gods and spirits, including animals and plants. We all have this medicine within us. It is a matter of awakening the divine spirit within to connect with the outer world.

In contrast to how expensive medicine, insurance, and drugs are, let alone hospitalization, traditional First Nations medicine men and women refused payment for their services; the power to heal was considered a gift from spirit, so they paid it forward. Otherwise, their medicine octave might sink into density and disappear. The principle remains: medicine loses healing power, creates a massive burnout culture, and loses its meaningful engagement when it becomes a commodity.

We may not commonly realize it, but animals are as important as any teacher, guide, or mentor that wants to share something with us. My editor Richard Grossinger wrote, "Some animals—insects, reptiles, mammals—have inborn medical intuitions, whether from DNA, hundredth monkeys, or morphic resonance."[1]

Animals also have oracular meanings; their presence and relation to your being in the moment has purposeful significance. In a divinatory sense like a tarot or oracle deck, seeing an animal emits a *human or angelic* frequency, or a way of resonating information to evoke a deep emotion, memory, or image that gently nudges us on our path. It depends on the animal; each has its own cosmic message. We may spend countless hours studying animal literature, classifications, and read numerous indigenous stories to better interpret the meaning of an animal. However, this method can lose the background of the animal in real time, the action of the animal, and its relationship with your life because of your common and divergent evolutions. Animals and humans share similar DNA, and they should not be discriminated against based on a hierarchical system of being more or less intelligent or dominant. We may not share furry or scaly physical properties, but our similarities can be shared in relationships, procreation, and survival mechanisms. My method is to cultivate your understanding of animals so that it helps, heals, and offers holistic support of your evolution and soul's purpose.

You are looking at a somewhat near DNA assemblage to your own.

Other animals are a gift from nature in their archetypal expression. Their greater identity isn't linked to one specific interpretation, and each of its archetypal elements reside in all of us. There cannot be a right or fixed meaning. When we embark on our own evolution, we are also redefining our relationship to nonhuman animals. They are no longer just a form of life that we share space with, they are ancestral support, teachers, and a moral compass.

Animals and plants are a part of us, essential to every breath. An animal, a holy being, isn't any different from our intrinsic connection to angels and our breath. *Anima* itself means "having breath." This approach heals our DNA's electromagnetic field and changes our body's signal. Spirit and totem animals interact with us on a higher frequency.

When you embark on this healing journey, you will find that animals start to naturally gravitate to you. That is normal. They pick up that frequency and are naturally more generous and clairsentient (able to perceive emotions and energy) than us in that regard. It's why spiritual teacher G. I. Gurdjieff urged students to practice love first on animals; he thought them more sensitive.[2] Establishing a dialogue with animals changes your internal dialogue with your DNA and its subtle configuration. It's like learning a new language: when you do, the size of the brain's hippocampus and cerebral cortex increases with it, strengthening its neuron connections over time. What happens with sentience happens with clairsentience, and it has the aura and upper chakras (throat, third eye, and crown) to work on its behalf.

The union of humans and animals has always been a holy, divine, divinatory experience. Ancient Egyptians likewise treated animals as gift-bearers from the gods and proxy humans. Pets were mummified in the same way that people were. Pythagoras believed that animals had the same type of soul as humans do. In non-Western tribes, animals are equal members of clans and sodalities as our brothers, sisters, aunts, and grandmas. Shamans incorporate their skins and skulls as robes and masks and imagine the spirit of the animal into their healing practices. San cultures in Africa dance in animal form to draw power for rainmaking, capturing game, and healing. For a similar reason, Asian martial-arts forms have animal names and choreographies: hawk, chicken, dove, turtle, monkey, horse, bear eagle, snake.[3] Practitioners of Xing Yi mimic the way a bird swoops, viper coils, rooster pecks, turtle snaps. They spread their "wings," plunge, pounce, shimmy side to side, and hop to close gaps. They ruffle their feathers and peck with a five-finger beak.[4]

I studied and played with animals from the time I learned how to walk and talk. My parents gave me the nickname "Matoaka" (who many know as Pocahontas, and originally Amonute) to represent my curiosity and barefooted playfulness in nature. The woods, plants and animals were my best and only friends. They didn't care that I couldn't speak English at the time or didn't fit into the ordinary example of what is

"cool" in culture. I came home on several occasions with ringworms from carrying feral cats, aching warts on my feet from barefoot hikes, and rocks and sticks as my primary childhood toys (I admit, I also had a beloved Barbie doll and a stuffed wolf animal).

When it was time to go home from school and my homework was already done, I knew my destiny would be the Appomattox River to learn more about water, insects, animals, and the sun until dinner time. They were my childhood, nurturing my heart with the wildest form of love. I heard their stories for longer than I seem to have been alive. I mean to honor their wisdom by passing on the messages they have passed to me.

I'm not here to tell you that learning animal medicine is an ultimate path to knowledge or evolution. I'm here to share how you can speak in an ancient language to your soul and reclaim a lost world of life, gifts, and understanding. By doing so, your mind is supporting and becoming one with all of life. But if we treat animals and plants as separate bodies from ours, we will stay on square one, indefinitely.

There are different methods of getting in sync with the animal kingdom. Some people spend many hours perfecting their *caws* and *ribbets* or doing monkey and bear Xing Yi or chi kung sets, while others assimilate animal lessons on the Discovery Channel and YouTube. I have my own method, a psychospiritual approach that encourages you to induce a relationship with the spirit of the animal through meditation and introspection. My approach is built on physical observation, empathy, and listening to frequency changes in my body, which I have developed through my personal practice of sound meditation. This approach takes time, practice, and patience, and it works. In the sense that I utilize this approach in my personal interactions, keeping the heart (or desire) and mind (the organized faculty of senses) open to learning and accepting. However, there are no fixed interpretations in the following examples, and the defining key is for your mind to accept fluidity and the nature of spontaneous change. Be respectful of all traditions and perspectives, each story has wisdom and ancestral density.

Though my descriptions of animals emphasize traditional as well as divinatory aspects, the key is not their actual characteristics, which may be common and well known, but the way you receive them like oracle cards. You may see a hummingbird, swan, or bear and look at it as a live form of nature, and that is part of its transmission and divination, but you can also look at it as a figure on a card drawn from a deck, the deck being the world and the reader being you. In that regard, finding a piece of cardboard with a swan image or a part of a puzzle piece with a fox in the dirt is not that different from seeing an actual animal, because you are the one who reads the oracle and assimilates the energy brought by the form. It is both a live biological entity and a petroglyph in your brain.

We will now explore the symbology of animals of air, water, fire, earth, and aether.

ANIMALS OF AIR

The energetics of air encompass intellect, inspiration, logic, memory, knowledge, comprehension, dreams, idealism, expansion, voice, and breath. The passage of breath allows intelligence to move throughout the body, enlisting the Higher Self. Each of these animals have, but are not limited to, a connection with the air element and the way they orient this wisdom in our lives. Evaluate the moment, be honest with yourself, and reflect within when you see our flying allies.

Hummingbird Essence
A Reminder to Look Up and Smile

Feeling the wakening of the sweltering morning, I watered rows of herbal arrangements of basil, rosemary, mint, eucalyptus, cilantro in the early dappled light of the nursery. In my mission to further my knowledge in the well-being of plants, working part time as annuals associate during 2023 was one of the best summertime memories of my life. This particular morning was special. Lowering a relaxed gaze, I listened to the continuous stream of the water and felt at one with the greenery that surrounded me. I consciously inhaled, breathing in this shared sophisticated hallmark of life and appreciated what seemed like a nested experience of unified intelligence with the trees, flowers, herbs, birds, butterflies, moths, bees, and all forms of life that live in this ecological balance of the gardens. In that moment, I felt my high ponytail vibrate, a playful pull, a sweet sentimental point of contact. Thinking it may be the wind playing with my hair, I thought nothing of it. Suddenly, my coworker and plant teacher Elaine Rush yelled my name. I looked over at her, seeing her hands over her mouth. She looked as if she saw a phenomenon. "There was a hummingbird circling around your hair like a halo! Did you feel it? It was there for a long time!" Hearing her validation and elation at experiencing the coalescing of nature firsthand, my feeling of awe and desire to extend myself and be a part of this larger system we know as Earth was fulfilled. The rest of my day was filled with nostalgic bliss. In remembering how I energetically shared myself with this ecology, I decided to share the essence of the hummingbird's gift of connection. By sharing my smile with the rest of my coworkers and the customers, I knew I could share and co-create happiness with them too.

The essence of hummingbird is persistent and social. When we see a hummingbird, we often stop what we are doing to watch it. That's partly because of the astonishingly rapid beat of its wings, but that beat and the birds themselves express joy. When they come into our lives, it's a good time to reflect on what you are grateful for, what brings you joy. When you see a hummingbird, do it. These fairy birds can often be a reminder to smile because their presence inspires awe and wonder at nature. Their presence is always welcomed. Hummingbirds in dreams can be an invitation to explore what is currently making your life colorful, no matter the circumstance, and support you in your capacity to pursue it. You've been chosen for a courageous role, and the hummingbird cheers you on.

Crow
Reflect on Change Beginning with the Past, Present, and Your Goals
Crows carry memories of the past, present, and, by their pattern of grouping—like a fortune from an oracle—can have a panoramic grasp

of what is next. It is the formula that makes up the nature of dreams. They remember the slightest detail and are quick to make decisions in the heat of the moment. Their medicine is being able to see the supernatural, the unseen and unknown, similarly to the way we could receive this information from our dreams.

Crows are thick with symbolism. They represent the infinite beyond the physical and spiritual world that humanity inhabits. They are reminders of the magic that we all carry—the ability to honor the past as a teacher, honor the present as creation, and honor the future as inspiration. Crows can mean change. They remind us to take a moment to reflect on our actions, past and current, and apply ourselves with confidence in addressing the issues that are causing a lack of harmony in our life. They are also psychopomps, guides to the afterlife and past lives. When you see a congress of crows, look through the lens of the past, present, and future—it is the gateway to the unknown that they provide.

Bee
The Fruitfulness from Hard Work and Artistic Expression

Bees are one of the hardest workers in the animal kingdom. Bees provide more than just honey. Bees pollinate apples, strawberries, onions, cabbage, avocados, broccoli, grapes, coffee, kale, cocoa, almonds, beets, lemons, lima beans, figs, kidney beans, cherries, green beans, celery, walnuts, pumpkins, flax, buckwheat, macadamia nuts, fennel, limes, carrots, persimmons, cucumbers, hazelnuts, cantaloupe, coriander, caraway, watermelon, coconut, tangerines, brazil nuts, mustard seed, cauliflower, turnips, Brussels sprouts, bok choy, papaya, safflower, sesame, eggplant, raspberries, blackberries, black-eyed peas, vanilla, cranberries, squash, mangos, kiwi fruit, plums, peaches, cashews—the list goes on. If the bees were to become extinct, much of the food chain for humans and animals would come to an end. The world would experience absolute scarcity. When I first started beekeeping, I brought ten thousand hard working female honey bees home from a local farm in the central mountains of Virginia. Only two weeks later, I was shocked and inspired to see they not only designed the honeycomb, but also created sweet liquid honey!

How many times have you seen bees working hard at the beginning of spring to help pollinate the land that supports all other ecosystems, including us? Have you ever paused to appreciate the sweetness of honey or beeswax and imbibed its health benefits? Have you stopped to thank a bee for its loyalty to hard work in such a short life span? Have you reflected on your own work and whether you are intentionally conserving your energy, organizing your time and discipline to engage with your set goals?

Bees model creativity, hard work, and fertility. Earth is the planet of work in this solar system. Everything gained here is by labor and practice. Bees are also team workers and depict the fulfillment of working as a group and individually. When you see a bee, think about an activity that is communal, meditative, and brings your creative muse delight. Explore it. Create your own joyful task, and afterward, take a step back to see how much you accomplished in a designated period of time.

Bees are also artists, able to understand subtlety. My partner plants exotic flowers, herbs, and vegetable gardens not only for me to enjoy, but for other plant and animal life. Bees have a way of taking over the garden through their own craft, making it their own. It's a phenomenon to watch.

Bees tell you to pollinate your own life.

Butterfly
Taking an Intentional Pause

When I was twenty years old, I took a summertime gig at the first annual butterfly exhibit at the Lewis Ginter Botanical Gardens in Richmond, Virginia as a butterfly technician. My role was to help import and settle exotic butterfly and moth species from India, Australia, China, and South America after they hatched from chrysalides. It was an incredibly fun use of my time to witness them hatching from a chrysalis in one moment, then taking flight the next. It is an exposition of birth. A scientist once said that you can nowhere find the butterfly in the worm that turns into it. It is invisible.

Butterflies are graceful reminders of transformative experiences and the lightness of being. For a short period of time, their chrysalis is a proxy for the vessel of our human body. They are testaments of our continual transformation in life. Floating gems of pixie-like magic, butterflies won't be contained, even in a butterfly form. When you see a butterfly enter your field, their awe and grace naturally forms its own dimension, and nothing else exists in the moment, except the butterfly. There is no other moment such as this one.

Thank it for the reminder to focus on the now, which acknowledges the next moment—that change is inevitable, constant, and as beautiful as it is. Feel that the path of form is a cyclic sequence of impermanence, and that is butterfly-like in and of itself.

Geese
Putting Your Foot Down

A flock of geese lands in the middle of an urban area and disrupts traffic. They cross streets and march through parks, joining homeless encampments. If you see these geese, you may be amused by them. They tromp along, chomping on grass and weeds and even trash. They cross a busy thoroughfare regardless of whether the light is red or green. What they *really* want to say is, "We are putting our feet down. We are unperturbed birds, so we have powers of air and ground together." If you happen to come across this display or draw a goose card from an animal deck, take inspiration from the geese's assertiveness and stay true to yourself, even in the face of adversity.

ANIMALS OF WATER

Water energy is healing, emotional, fertile, intuitive, sensitive, deep, lunar, and renewing. Bodies of water can be an emotionally nurturing environment and also powerfully destructive. Animals of water continue to bring life into springs, streams, seas, and oceans that have gone through subsequent renewal cycles. This concept alone carries the evidence that life continues to reemerge in the repetition of life, death, and rebirth, In examining the life cycles of these water animals and their connection to elemental medicine, we can gain a deeper understanding of their symbolic meaning and wisdom they offer as oracles.

Turtle
Reminder of the Rewards of Stillness

One of the oldest First Nations tales is that of the beloved turtle, shifting by tribe. In her book *Braiding Sweetgrass*, citizen Potawatomi Kimmerer shares the account of Skywoman, a star in the constellation of teachings called, in translation, the Original Instructions, which are ethical prescriptions for respectful hunting, family life, and ceremonies. As Skywoman fell from the sky, she found herself floating in water on Earth among geese, otters, swans, beavers, and fish. A turtle floated up to her and offered its back for her to rest upon. As the animals offered her a place where she could live, she sang in gratitude, dancing, and letting her feet caress the land. The land grew as she danced and gave thanks with the alchemized animals' gifts. Together they formed Turtle Island.

When I run to my local park, I enjoy seeing a group of turtles huddled around on a wide stone in the middle of the lake, if the water levels are low enough. I refer to it as the "Turtle Island" mentioned in *Braiding Sweetgrass*. There's a sense of divine serenity in the way they sit together. Suddenly, I see one turtle snap at a fly. The rest of the turtles remain still.

At another lake in my city where people fish and enjoy watching hundreds of turtles in the evening, thousands of fish died overnight. The catastrophe was caused by overheating, lowering levels of dissolved oxygen. What happened to all of the dead fish so that the lake didn't become contaminated for other critical ecosystems? The turtles ate them.

Turtles remind me of the power of stillness, waiting, and quietness. It takes a lot of patience, trust, and skill to be calm. It also takes a lot of courage and determination to intentionally slow down considering the "normal" neurotic tempo of the modern world. Turtles pick up the debris of the deep. They remind you to go deep into your own ocean. They are marvelous and strange things that attach to your shell. You can't see them, but they work their way into your mind. However, the rewards of patience are endless. Ask yourself what can you extend your patience to in your life? What do you anticipate from such stillness?

Frog
Listen to Your Body When It Seeks Cleansing

Frogs have a diverse array of water aspects relevant to cleansing. Frogs represent purity and replenishment by addressing negative energy, filling your body with nourishing energy when it feels dried out, and cleansing our soul by honoring tears. Frogs call to the universe for all of us.

Kambo is a healing ritual in South America by traditional indigenous groups such as the Noke Kuin, also known as the Katukina, in the Brazilian Amazon rainforest. Using the poisonous secretions of the giant monkey frog, a small dose is applied to a small burn on the skin where it enters the lymphatic system through the bloodstream. When I first experimented with this ritual, I felt frog DNA *scanning* my body for problems. The first ten to fifteen minutes were unpleasant, but it addressed exactly what my body needed. Why would anyone engage in such an amphibious process? Usually, our minds are either too intelligent or stubborn to allow us to purge and heal. It takes the unconscious to acknowledge where a certain part of our body needs healing. When kambo is induced, your body's response goes into a, "Oooh, I had no clue this trauma was sitting here all this time." The frog serum allows the unconscious to be revealed. So, I allowed it to open my heart, feel the emotions, shake it out like a lion, and experience the bliss of a physical and spiritual release.

A frog's *ribbet* or *coqui* isn't only a call to the wild for mating or bringing in rain, it is also to remind us of the constant purification of life. If we seek healing, our pursuit must be proactive. When you see a frog, ask yourself where you need cleansing the most. Maybe it could mean a relaxing salt bath, listening to healing sounds, disconnecting from social media or an addictive habit, or simply taking in deep, slow breaths. It could also be a *ribbet* of forgiveness. Is there someone in your life, even yourself, that could be forgiven?

Salmon (or any Fish)
The Gift of Life

Salmon live to spawn. Once the female salmon cover the gravel nest that she has prepared with thousands of eggs, her body dies and becomes nutrients for the ecosystem. In Vancouver, Canada, millions of salmon arrive at the rivers and lay eggs during the fall season. More than a hundred different types of animals feed on the salmon and their eggs to prepare for the winter; it is one of the largest migrations on the planet. Bears carry thousands of pounds of salmon into the forest. Up to 80 percent of the nitrogen in the trees comes from the body of the salmon.[5] Their carcasses go back to the sea and become a feast for the eagles. The salmon not only feeds marine and terrestrial organisms, it gives life to the trees, the soil, and the inland island.

Similarly, the herring lays eggs and creates a jewel-studded veil of new life: half a million eggs per square foot.

Take a moment to contemplate the lesson we can learn from the salmon's life cycle. Just as the salmon's seemingly small act of laying eggs has a profound impact on the ecosystem, consider how your actions, no matter how small they seem, can contribute to your personal growth and the lives of those around you. Reflect on the "eggs" you are laying in your life and how they may shape your future.

Swan
Trusting Your Sense of Intuition

Swans are a master of lightness. Typically, they stand for grace and elegance, but not only in movement or appearance; the swan reminds us of identity. It is a reflection of trusting your intuition—one of our life's greatest challenges.

How do you surrender to life and meaning? Are you willing to accept healing at this moment? How do you feel about change? What are the things that you are willing to see change in your life? How do you practice flow in your life? When you see a swan, they may be hinting for you to accept or acknowledge a transformation, and gently reminding you to remember your innate grace. In other words, be you and flow into it.

Otter
A Tough Nurturing

A female otter and her young spend about six months to a year together.[6] During this time, the mother constantly feeds her young and herself, ensures her pup is safely floating on top of her, and trains it how to source its food. She is exhausted but continues to nurture her young solely. The father does not participate in the pup's development. When the pup is about a year old, the father returns for another season of mating. The pup is instantly and abruptly cut off from the mother. The father "reclaims" his territory, and prohibits the pup from coming back to it.

The otter's life serves as a reminder of the importance of accepting life's challenges and acknowledging the role that tough nurturing plays in our personal growth. The mother executes her role in preparing the pup's life until the moment the pup *must* go. That is an aspect of nature.

Have you experienced a challenging moment, possibly leaving you with grief and where you must grow? Otter medicine is powerful and, in a sense, dramatic. Oppositions are similar to cycles; both are necessary to create a fully rounded experience for evolution to take place. What can you let go from a previous chapter in your life? What are your ambitions for the next one? What can you thank your parents for to continue in your path of creation and letting go?

ANIMALS OF FIRE

Fire energy is ardent, enthusiastic, spontaneous, self-sufficient, courageous, purifying, lusty, clashing, energizing, romantic, and it will become your teacher. It governs all transformation. Animals of fire teach us the power of tinder through mastery of their own skills. The element of fire serves a common athanor, or alchemical furnace, that opens doors to the spiritual world.

Lion
Observation Is Action
The lion is traditionally described as "stalking" its prey, but that is a false human perception. Lions observe and track movements and changes. They don't necessarily attack or strike, unless they are doing what nature naturally requires all of us to do—survive. That is the power of the lion. It is not its "royal" place in a made-up hierarchical

system of "animal society." The lion is a co-participant in the animal kingdom. Its role *is* to observe and match its unparalleled strength with an innate responsibility to share the terrain.

Perhaps the distinction of power and strength in life is not emphasized in its physical expression. Maybe it is the pause, hesitancy, or courage to hold back in certain times (or most times). Maybe it is simply listening to animal languages to get to know other species without an alternative agenda, such as survival (or trade, for humans). Or possibly, it is the general act of participating in life—to be in awe of the frame before you.

Lion medicine is potent. Ask yourself: what do you seek to understand or want to become a part of? Sit with these questions, literally, metaphorically, spiritually. What shifts do you feel within your body or in your perception as you take on an observing role? How do you share space with others in your life? What are the types of things that you wish to feel ownership of?

Fox
Intentional Humor to Blend In

The term *fox* has many connotations in the English language. It is usually associated with slyness, but I find this an oversimplification. The magic of the fox is in its ability to camouflage. Camouflage is an intrinsic aspect of nature and developed by almost every animal in its evolution. Chameleons, insects, owls, octopuses, moths, crabs, geckos provide examples. It is a gift to meld into the scenery of landscape and become invisible to observing predators. The family of foxes that live in my backyard feel safe enough to make themselves be seen, but otherwise, they are a rare sight.

I am fascinated by humor as a tool and intelligence. Humor blends a frame of reality by a cognitive phenomenon. A verbal exchange occurs between two people for relevant information about a particular subject. The space in between a conversation is structured by elements. The elements pertain to the particular scenario of the relation between the speakers, but humor can be an element as a means of comprehension, connection, and sometimes, even a way of retrieving information. As

Paul Lewis writes, "humor embodies values not by virtue of its content alone but as a consequence of what it does with its materials. To get a joke we must resolve its incongruity by retrieving or discovering an image or idea that can connect its oddly associated ideas or images."[7] When I recognize "foxy" tendencies through an unusual or foreign joke and sense it is a facade, I ask the questions or make a comment such as, "Why are you laughing? What is funny? I do not understand the joke." I listen to the response to clear uncertainty or expose insincerity.

Foxes use their craft in camouflage through tactics as a maneuver. Integrating humor into conversation is a cunning way to observe another person's thoughts and behavior. Be sly and weird, but be careful with this tactic: it can burn a bridge in a relationship or create a rather uncomfortable situation for a conversation when fox elements are recognized.

Snake
Alchemy, Transmutation, Composting

The snake is a messenger of healing, fertility, desire, and creativity in the world. It has traditionally been associated with the water element as embodying feminine energy, Shakti, or Kundalini. I've classified the snake with fire because its element of transmutation resonates with a capacity for fission each person carries.

The alchemy of transmutation gives us the power to heal emotional wounds. The lesson of the snake is that you have the ability to heal ancestral trauma and grief and regenerate new life and vitality. This level of medicine is universal and whole, beginning with your state of consciousness.

If you see a snake, it could be time to ask yourself: What emotion or thought needs to be transmuted? How do you invite the power of fire into your life to activate yourself? Do you live with a snakeskin on, metaphorically? If so, why? How often do you transmute grief? How do you purify your connection with your ancestors to heal yourself, the past, and your future (crow as well as snake medicine)? How do you orient yourself to mindful healing? How do you become a natural ground healer? Remember too that snake venom is its own medicine. You contain serums in your immune system.

Coyote
Check Yourself!

On our way to Page, Arizona, for a cross-country road trip in 2022, my partner and I stopped at a remote gas station in a desertlike scene. Sitting in the car with the door open looking at the mountains from afar, we see a coyote jump out of the bush about twenty feet away, doing a circle dance with a serious, but playful look. After a couple of seconds of what felt like eternity by the startling surprise, it ran off. "What the heck was that?" I said to my partner, laughing. It was a whimsical oracle.

Coyotes are tricksters, for life. They have a lot of medicine powers, but they don't necessarily use them. Coyotes are elusive and intelligent but are so tricky they can also trick themselves. They are great at repeating the same mistakes, without registering any lessons. Coyotes are self-con artists and their foolish antics can be amusing to watch.

Coyotes show an aspect of ourselves that exists in everyone. It is an easy thing to slip into: a state of self-enthusiastic talk but no proper action or experience to show. I've known two people who have embodied their focus and passion in life, my dad and Richard. My dad woke up at four each morning to get to work and wouldn't be home until six or seven in the evening. He would work double shifts at times or pick up some extra shifts as a pizza-delivery driver when I was in middle school to make financial ends meet. He never talked about how hard he worked. For Richard Grossinger, do a search on the variety of books he has authored and published in his lifetime—and still counting. Their actions, discipline, and commitment exemplify the action-oriented approach embodied by coyote medicine, inspiring us to transform our intentions into tangible accomplishments. Similarly, a coyote can bark at a tree sometimes for as long as an hour because it wants the squirrel in sight to come down—but we all know the coyote's aimless singing won't help. In both examples, coyote medicine demands action.

Coyote medicine activates what is under the surface. It's an itch to change former ways that lead to dead ends. What are you willing to sacrifice to focus on your goals? What is one thing that you can organize in your life to help you focus on the topic you've been talking about but haven't gotten around to doing yet? What will encourage you to take action? When you get around to it, you won't be laughing or talking about it anymore. You'll be doing it.

Firefly
Dynamic and Subtle Inspirations

At the brink of summer, the sighting of the first firefly is evocative and nostalgic. The firefly reminds us of the pulsating, phosphorescent source of inspiration and its orientation to impermanence. It is the same feeling when we are moved by a song, poem, photograph, dance . . . and love.

Fireflies are oracles of precious, fleeting life. Firefly medicine provides a way of looking at your own source of inspiration and seeking ways to support it. Our light shines the way for others, so that they can shine theirs. Together, we can illuminate the way forward with our generational evolution. Ask yourself: If we had only so many opportunities to shine our light for a short period of time, what would we do with it?

ANIMALS OF EARTH

Earth energy is life-giving, grounding, stable, dependable, conservative, fertile, death and rebirth, prosperous, wealthy, and sensual. After working with the element of water and emotional release, working with earth is helpful to become grounded and centered. The animals of earth offer their grounding medicine, a subtle way of cultivating stability, connection, and a deep sense of belonging.

Bear
Acquiring the Strength to Go All In

A sighting of a bear might make a person initially feel uncomfortable, but it also gives a sense of reverence. The strength of a bear is its own innate power. A bear acquires strength during the fall months, it prepares for its hibernation by eating and drinking continuously. Its long coma is not only important to its survival but beautifully woven into subconscious introspection. After the bear awakens from its hibernation, it begins a new state of life and direction. It enters a cycle of full nourishment, then goes into the void to dream and awakens to a new life. It teaches us that the cave, or the "void" in our lives, can be a surprising place to acquire strength. The question is, how brave are you to take a chance on going all in to the void to rediscover this kind of strength?

When I visited Abingdon, Virginia, to focus on my writing, I stayed in a cottage surrounded by four hundred acres of trees. Traveling alone, I trusted the woods and its wildlife. One morning as I walked to my car, a few-hundred-pound black bear was meandering on site, and we found each other with only a thirty-foot separation of air. I had nothing on me but a distinct pulsating heartbeat. It came alone, with no cubs. The bear and I had a stare-down for about ten seconds. It lifted itself on its back paws and continued its intense stare standing up. We were mirroring each other vertically, a woman and a black bear with nothing but space in between.

In that moment, I connected deeply with its inner trance. We became one in that specific moment. Besides feeling shock and uncertainty about what will happen next, the bear's piercing, eternally dark eyes asked me, "How far are you willing to go into the void to discover what is waiting for you?" After what felt like eternity, it dropped back on all fours, grumbled something, and slowly walked away.

Are you willing to explore the unknown aspects of yourself and uncover the strength that lies within?

Bison or Buffalo
As Fearless as a Shaman Warrior

I was about a thousand feet from a wild colony of bison the first night that I spent in the Badlands, South Dakota. Seventy-seven grazed the field (I counted), as my partner and I set up our tent and watched them preparing for their rest in the sunset. Some rolled around like playful dogs in the grass. A mother nurtured and groomed her calf and listened to its grunts and snorts.

Though the traditional accounts of bison and buffalo highlight the realm of abundance from the stories of the Lakota tradition, I see them having a different meaning of abundance. These enormous creatures have no fear of death. They see any hardship as an opportunity. They are shaman warriors. Their eyes remain ahead of the road. That is the meaning of true abundance. When we have such a strong mindset, we progress with gratitude. Ask yourself: What are you currently experiencing that brings you fear? What can you learn from the experience? Are you willing to take a chance and a step into that direction?

May we all experience courage for taking on our fears in life. May we treat life with gratitude and appreciate the way it shows up like bison in the occasional stampeding herd of our evolution. May we stampede toward our soul's plan like buffalo on the prairie.

Dog or Wolf
Territorial

Have you ever noticed the way dogs behave happily with their owner and family, but when a stranger is introduced, their orientation changes drastically? If you're lucky, they pause to sniff you first. Dogs and wolves are naturally territorial over their space and their pack. Wolf howls assemble the pack, and they are excellent communicators through vocalizations such as growls, whines, barks, and even heavy, deep sighs.

Dog or wolf medicine isn't so much about recognizing territory or space but respecting it. For example, when you invite a guest over to your home, there's a level of understanding that encompasses courtesy, recognition, and honoring to preserve the warmth, safety, and structure of your home or body-mind temple. A dog or wolf's home is its structure. Similarly, we can ask the question about our "territory" as in, how are we honoring and respecting the space? What can we improve on? Is there a necessary shift that is sparking your curiosity?

Deer
Gentle Spirits like a Subtle Breeze

I met a man named Bright Sky. Akin to a deer's intuition, Bright Sky possessed an uncanny connection to nature, evidenced by his remarkable ability to predict wind patterns with astonishing precision. "It is about to happen now," and as he said "now," the trees around us would start to sway back and forth. His connection to nature, similar to the heightened awareness of a deer, was mesmerizing.

Do you remember the first time you saw a fawn or deer? What did you feel? Did it remind you of grace and nurturing? A deer's charm is mesmerizing. They are the emblem of softness and gentleness but also mindfulness. Deer medicine reminds me of my grandmother, specifically the way she expressed her opinions, gave advice, and provided support. Wisdom keepers have heightened awareness and can melt the heart of any wounded creature. She always helped me find my path by simply being sweet.

When you see a deer, it is showing you the gentle mindful path. The occurrence is as subtle as a light breeze, but did you notice it? Either you are on the right one, or it is reminding you to receive the beauty of compassion and the gentle medicine of nourishing the world. Let deer medicine be a resident of your heart and mind, always.

Marmot
Adventures, Community, Wild Courtship

Marmots remind one of hobbits. Marmots are relatives of squirrels (just as hobbits are relatives of the human race), live underground, and seem to go on epic adventures. Marmots not only depend on each other for survival, communicating through piercing blasts of chirps, but also live in polygynous groups. As we were passing through the Rocky Mountains, we hiked to the top of a smaller mountain to see that we weren't the only participants at a gorgeous sunset. A marmot was staring out at the breathtaking landscape. It was probably scanning the area for any predators before it could safely find food, but in the moment that I took out my camera, it turned its head and gaped at me—almost as if it was saying "cheese!"

Marmot medicine is rounded and has multiple meanings. When you see a marmot oracle, even if it is its relative squirrel, ask yourself what are you in need of harvesting in your life? Are you seeking an epic adventure to experience? Are you longing for connection? Are you a wise Fool like the tarot Jester, alert to predators of air?

ANIMALS OF AETHER

Aether presents itself from beyond the realms of perceived reality. It is an element that represents mental space, spirit, intuition, stillness, light, sound, ancestry, love, and evolution. Aether cannot be created or manufactured by humanity, as it always exists in multiple realms and dimensions. The animals of aether can be seen as a connection to these higher realms of consciousness. While this is a limited list, it is safe to say that all animals can be associated with aether in the context of sharing relational information.

Eagle
Stand Up for Yourself
Touching the light of the sky itself, eagles soar high. Perching at the peak of any tree, they exhibit aims and capabilities that encompass the entire field. Even with their agility, speed, and strength, they sit and watch, patiently. They are not intimidated by anything of land or air. An eagle continues on its fated mission and accelerates through challenges. A mix of fire and air makes up their natural leadership qualities. They demonstrate the divinity of life.

When a young eagle learns how to fly and fish, it is common for another eagle to function as a competitor for its game or food. An experienced eagle will try to steal or push the young eagle away from its meal. At some point, the young eagle has to stand up for itself. In the eagle world, it pays to be bold.

Eagle medicine reminds us that we are capable of doing *anything* and will not be controlled, bullied, or settle for another competitor taking our meals! Imagine the doors you can open when you mix eagle and bison medicine together. The eagle element encourages you to embrace your power and evolution. How can you gracefully stand up for yourself in your life?

Hawk
Pay Attention to Signals
The Hawk also watches animal life from high above. Its ability to detect any subtle detail, movement, and life heat is like a superpower. Hawks, for being so feathered and physical, are messengers of the spiritual world, its incarnation in body. Quick and nimble, a hawk moves in the wind with purpose. Wind, a movement of air, also means news. When you see a hawk, it may remind you to take a step back and bring

awareness to all the other signs, symbols, and subtleties in life. Even a small message can help you shift the direction of your course.

In the space that you're currently in, take a pause. Relax your gaze and allow yourself to look around slowly. Settle your eyes on something that feels familiar and right to you. What does the symbol that you're looking at mean to you? What does it remind you of? How does it relate to your current life? What about your memories?

Dragonfly
Messengers of Illusion

I witnessed a remarkable dragonfly phenomenon. The same day that I visited the lake where thousands of fish were dead and eaten by turtles, I stopped in my tracks at the next sighting. At an area where the lake water drains, covered by a row of metal bars, on top of the bars, I saw four dead fish, one of which was a three-foot carp. *Hundreds* of dragonflies were flying directly on top of them without sitting on them. I watched to see if they would land on the fish, but they didn't. In fact, dragonflies are afraid of fish.

Dragonflies are messengers of illusion. While dragonflies are active, they represent the transcending spirit. When you see a dragonfly, you are being visited by a spirit. They reveal the evolution of its spirit, too. Their presence reassures us that the souls of our loved ones or other animals are alive and transformed into the spirit world. Similar to the dragonfly perching on my dad's tombstone, or the hundreds of dragonflies that symbolized the many fish in the lake, other dragonflies are spiritual shape-shifters and flash colors that cannot be found in our world in the iridescence of their wings. When you see a dragonfly, it could be reminding you that things are never what they appear on the surface. It is a gift to see a dragonfly. Remember to give thanks when you do. A spirit could be hitching a ride or it may just be carrying a spirit message.

Dragon
Keepers of Metamorphosis

Dragons are real. They are part of fire, part of air, part of aether, and also a part of a zone beyond earth itself. They come and go, and are found between the planes of consciousness and unconsciousness. They come in different forms, shapes, and colors. Some have tails, others do not. Some breathe fire, others can talk directly with you, some when you are awake, others in your sleep. Some prefer that you avoid touching their tail, others may invite you to jump on their backs for a ride. Some may take you back on memory lane, others may have a more serious mission of divination. They hide in clouds, air currents, waves, lightning, and other electrical phenomena. Dragons have been depicted culturally as divine warriors, and that they are. They are messengers, and their relationship to humans can come in the same astral and aetheric frequencies as angels. They can be protective, insightful, and a channel to your inner world's wisdom with a mysterious pulse from the aether.

When you see a dragon in your field, whether it's in a meditation, dream, trancelike state, imagination, picture, or shamanic journey, honor your time together by being as present as you can. Are you wishing to change gears in your life or reorganize your thoughts? Or update your belief systems? Are you willing to be open-minded to other frequencies surrounding you like the angelic realm or animal guides? Do you want to evolve?

Vulture
To Spark Purification and Composting

I was given a vulture divination card on three separate occasions by my partner nearly a year ago. It reflects the spirit that works within me. These scavengers, particularly the Rüppell's griffon vulture, which ranges across eastern Africa and the Sahel, are the highest-flying birds and can soar up to 37,000 feet above sea level, which is 444,000 inches, both pertinent values for my numerical approach that connects animals, angels, and divine gematria (the numbers 37 and 444 will be discussed in chapter 14.

Vultures are essential to the regulation of ecosystems, composting and purifying or cleaning up the planet, playing a major role in death and rebirth of life. They are one of the most undervalued, depreciated organisms. Their behavior is also intriguing to me; they are able to spot a three-foot carcass from four miles away; then they circle around it to draw attention to other vultures to join in. The vulture is the spirit of this book, drawing attention to the twelve strands.

To experience their medicine, try this: Pick an activity that requires motion. For example, washing the dishes, going for a walk or run, cooking, cleaning your space. Allow your mind to tend to what you're doing, but also relax. When you get into a comfortable flow, pay attention to the first constructive thought that emerges outside of the task that you're doing. Become curious about it. Don't let it consume you, but pin it in your local memory and allow life to reveal more information about it with time. Vulture medicine gears us to soar the psychosphere as a means to enter a meditative state of concentration and experience the sharpening of our focus to resonating thoughts and revelations.

REAL-TIME INTERPRETATIONS

Imagine yourself coming to your elder to ask for wisdom, support, or advice. When you approach your elder, you may feel intimidated. Because your elder has experienced generations of wisdom from trials, you care about what they have to share with you. How would you address an admired person in your life? Would you approach them from a place of humility? Would you give your undivided time, presence, and attention to them? How would you treat a loved one knowing the limitations of their time and yours? All of these questions can help you prime the way you connect with an animal or entity that will share its meaning with you.

There is an elder that lives within you and an elder that lives within an animal. When you meet an animal, these elders can talk with each other. It's an intuitive language, without identification of "yes or no," "right or wrong," "you or me." Their exchange is the heir to "ours, us, and we." With reverence for both, the conversation becomes directional in both value and harmony. It doesn't mean that you can now just situate yourself in dangerous situations with wildlife, but it is a way you can learn how to balance spirituality with animals. In a dream, I was startled to experience a black panther swimming toward me, showing its gnarly white teeth, as I felt cornered into a tree. The dream also takes a surprising U-turn when the panther jumps on the tree behind me to growl at any opposing threat to me. The dream could be depicting protection and divine support in future endeavors. The panther was my ally, not my threat. Some people go to sleep at night in the wilderness and wake to find a bobcat or lynx licking them. From that point on, they are under the animal's protection.

There aren't any books or guidelines on how to accurately interpret animals and dreams. However, if we take the approach of asking the animal what it would like to share, we usually do come up with an answer. In time, the result shifts and evolves, and so do we. We become better interpreters and speakers of this language. My journey entailed thirty years of watching and listening. When I started to seek a conversation with my animal ancestors, I received responses because I cared to engage in their language.

The next time you see a spider, ant, cat, dog, deer, rabbit, frog, butterfly, moth, bee, earthworm, beetle, or anything, tune in to the moment and ask both of you (you and it), "What are you trying to share with me at this point in this place? How can I give back to support you?"

Divine Codes, Human Origins, and Angelic Frequencies

When I talk about frequencies of creation and nonphysical forms, I'm drawing on traditional esoteric and mythological references as well as scientific advancements in microbiology that followed the discovery of the DNA twin helices. For DNA, we have electron microscopes and lab experiments. When aligning with angelic energy, my inspiration originates primarily from my deep thought, meditation, and life experiences. Each is necessary for this vision. With this in mind, let's explore the origins of our humanity as it was channeled into the physical plane from the celestial.

THE GEMATRIA OF ADAM AND EVE

In Genesis 1, "In the beginning, Elohim created the heavens and the earth" depicts the spiritual creation of the Universe, not just one planet. Genesis 2 shows the second creation, the manifestation of spirit into reality as humanity: the first angels became humans. The dichotomy between humans and angels is equivalent to the masculine and feminine or two separate beings of any sort or gender. The relationship between Adam and Eve symbolizes the partnership between humans and angels, regardless of gender.

Earlier, I discussed the significance of numbers as vibrations. In the angelic frequency, numbers are treated similarly, but there's a special meaning behind the numbers 1 and 0. In Jewish mysticism, zero is a nonbeing containing unlimited powers, and one is the beginning of a physical state. Zero relates to a spiritual being, and one is a physical being. If they are treated separately, they don't have meaning other than as a single digit vibration. However, when they are coupled together as integers, they form 10. Ten as a binary system not only makes up our modern technology, but also ten is a number of creation, which includes the angelic realm. Migene González-Wippler equates the ten fingers that human beings have with the ten cosmic laws, the ten angels, and the ten spheres of the Tree of Life in his *The Kabbalah and Magic of Angels*.[1] In *The Kabbalah and Magic of Angels*, Migene writes, "It is of transcendental importance that we understand the meaning of the binary system, because upon the number ten hinges everything that exists and everything that we desire to accomplish."

Let's look at the gematria relationship between Adam and Eve in Greek (in bold), Latin, and English:

$$\mathbf{A\Delta AM} = \mathbf{46}, \text{ATHAM, Adam}$$
$$\mathbf{EYA} = \mathbf{406}, \text{EVA, Eve}$$

The zero in Eve seems to represent both the angelic counterpart in humans and the female counterpart to Adam. In this gematria sum, we observe the angelic and spiritual relationship of angels and humans. The humans were modeled from the design of angelic beings, which primarily includes the spiritual essence of masculine and feminine in union. In a meaningful coincidence, the number for Adam also reflects the total number of human chromosomes.

The phrase "the spiritual energy" has the gematria sum 1101 in Hellenic, Η ΠΝΕΥΜΑΤΙΚΗ ΕΝΕΡΓΕΙΑ (the Latin form is I PNEVMATIKI ENERGEIA). We can observe the synchronicity in the zero between the three ciphers "111" that also shows spirituality. Recall that in chapter 9, 111 Hz is the sacred vibration that was used in the Hypogeum of Ħal Saflieni, the holy underground temple where the ancients used to heal. The word θεμελιώδης, meaning "foundational" or the basis on which something is established at the beginning, has the gematria sum 1111.

THE GEMATRIA OF GOD

My investigation led to another interesting observation of the mystical relationship between the divine and geometry, which involves two numbers that are interlinked: 144 and 37. In chapter 17, the Greek gematria for the word Θειον, or divine, is 144. The number 37 has a number of interesting associations. It represents the distance in light-years from Earth to Arcturus, a significant star in the constellation Boötes. Our mitochondria contain 37 genes. Notably, the popular yellow capsule-shaped minions from the *Despicable Me* franchise also measure 3 feet 7 inches in height. Furthermore, number 37 symbolizes divine unity, as its pythmentical (*pythmentical* refers to the single digit of the sum of the total number) value is 1 (3 + 7 = 10 = 1 + 0 = 1). The numbers 144 and 37 both share the twelfth placement in the Fibonacci (F0 = 0 and F1 = 1) and prime number set sequences, respectively. Fibonacci sequence can start with either 0 or 1, based on the specific context or preference. Usually, it begins with 0 followed by 1. If you start with 0, it is important to remember not to mistakenly count 0 as the first position. It should be labeled as F0= 0 and F1 = 1. Mislabeling the starting position of 0 as 1 will misalign the numbers. For example, the 12th number in the sequence wouldn't correctly correspond to 144. This alignment is key to maintaining the sequence's accuracy.

A mysterious number is encrypted in an equation of science and spirituality. Human DNA can be symbolically represented as the property of divinity (144) multiplied by (×) unity (37), signfying a harmonious connection with the Universe and within oneself. The common denominator of 12 strands of DNA represents the dodecahedral molecule and the placement of the numbers 144 and 37 in both Fibonacci and prime number sequences. By treating 12 as a dividend of human DNA, its numerical component can be decrypted:

$$\text{Numerical component to be decrypted} = \frac{(144 \times 37)}{12}$$

The result of the equation is 444. Before we get to my analysis, I want to discuss elements of the work of Italian music theorist Guido of Arezzo. Guido was a Benedictine monk who played an influential role in establishing a hexachord scale (a six-note musical system) to create new sound in the twelfth century, which led to medieval and Renaissance music. His hexachord system, known as the "Guidonian hand," introduced the solfege notes (ut, re, mi, fa, sol, and la) that became a foundation for music learning, paving the way for modern music and revolutionizing music education. Principally, the Guidonian hand was a simple tool that helped musicians, monks, and nuns remember the notes and sight read music. Beginning at the thumb of the left hand, the lowest note begins at Gamma *ut* (tip of the thumb) and progresses down the joints of the thumb followed by A *re* and B *mi*. The remaining notes follow a spiral pattern around the hand, moving up the pinky, across the fingers, and circling back. If we treated the Guidonian hand as an ancient English alphanumeric system—instead of using the Greek one—with the letter *a* corresponding to 6, *b* to 12, *c* to 18, *d* to 24, and so on, the gematria for the number 444 becomes intriguingly different. Using this system, the phrase *code of God* shares the same gematria sum of 444, as does the word *gematria* in English.

PLATO, FIBONACCI, ANGELS, MORPHOGENESIS, AND THE TETRAGRAMMATON

In different chapters of this book, we examined how geometry, angelic frequency, morphogenesis, numbers, sounds, and words have a symbiotic relationship to the 12-stranded DNA. This is a remarkable coincidence. Each concept shares an equivalent gematria sum: Plato, Fibonacci, Arcturus, morphogenesis, and the Tetragrammaton create a synapse by their gematria sum, 1261. It is arbitrary on one level, and on another, it is the root that holds together and communicates reality on this plane. The pythmentical value is 1, which represents unity:

> **Hellenic: ΠΛΑΤΩΝ = 1261**
> Latin: PLATON
> English: Plato
> **Hellenic: ΦΙΜΠΟΝΑΤΣΙ = 1261**
> Latin: FIMPONATSI

English: Fibonacci
Hellenic: ΑΡΚΤΟΥΡΟΣ = 1261
Latin: ARKTOUROS
English: Arcturus
Hellenic: Η ΜΟΡΦΟΓΕΝΕΣΙΣ = 1261
Latin: I MORFOGENESIS
English: The Morphogenesis
Hellenic: ΤΕΤΡΑΓΡΑΜΜΑΤΟ = 1261
Latin: TETRAGRAMATO
English: Tetragrammaton

$$1 + 2 + 6 + 1 = 10 = 1 + 0 = 1$$

Plato and the Five Unique Solids

At the basis of regular geometry are the five unique solids introduced by Plato, in his *Timaeus*. Plato, along with Fibonacci, tried to decipher the mathematical order of nature. He discovered that cosmic energy produces a non-acoustic sound and imitated this rhythm to find hidden levels of significance and secret codes. The use of the term rhythm references the attempt to explain order of the veiled patterns in nature. Each of Plato's thirty-six works* is consistent in respecting twelve as an important number. His productive uses and forms of dialectic inspired me to integrate gematria to introduce the 12-stranded DNA using numerical and etymological frequency.

Let's look at the significance of Plato's five solids. Five is a prime, a whole number that cannot be written as the product of two smaller numbers. The prime number sequence is 2, 3, 5, 7, 11, 13, 17, 19, 23, 29, 31, 37, and so on. Prime numbers are special in number theory and encryption. In ancient times they were even associated with the supernatural. The astronomer and science author Carl Sagan wrote a book

---

*The thirty-six works of Plato are Euthyphro, Apology, Crito, Phaedo, Cratylus, Theaetetus, First Alcibiades (Primer Alcibíades), Second Alcibiades, Statesman, Parmenides, Philebus, Theages, Sophist, The Symposium, Phaedrus, Minos, Hipparchus, Erastai, Charmides, Laches, Lysis, Ethydemus, Protagoras, Gorgias, Meno, Epistles, Clitophon, Hippias Minor, Hippias Major, Ion, Menexenus, Republic, Timaeus, Critias, Nomoi, Epinomis.

called *Contact* in 1985 about extraterrestrials trying to communicate with human beings using prime numbers as signals. Metaphorically, they can be likened to God, in the way that the number exists by itself and cannot be reduced other than by 1 and itself (e.g., the factors of 37 can be given as 1 and 37), exemplifying sacred duality. The paradox with prime numbers is that there is an infinite amount, yet they become rarer as they become larger.

But there is more to five-ness. There are five universal elements, fire, earth, water, air, and aether. A pentagon has five sides. The gematria sum of the Greek words *regular pentagon* (Κανονικον πενταγωνον, 1750) is equivalent to the sum of the words *divine archetype* (Θειον αρχετυπον, 1750), possibly referencing an attribute of the angelic realm.

Fibonacci

Everything in the Universe can be measured from the human body and energy fields around it through geometry. By closely examining geometric shapes and proportions, we know the basis of sacred geometry is the golden ratio or golden mean, which was first defined by Euclid (discussed in chapter 7). Fifteen hundred years later, Leonardo of Pisa, commonly referred to as Leonardo Fibonacci, introduced a numerical sequence that results in the golden ratio. The golden ratio, approximately 1.618, emerges upon division of a Fibonacci number by its predecessor. Ironically, Fibonacci never learned the sequence's most significant characteristic is its convergence of successive members of constant phi.

Known since antiquity, subsequent numbers of the Fibonacci sequence are added together to arrive at the next term (0, 1, 1, 2, 3, 5, 8, 13, 21, 34, and so on). The Fibonacci sequence has a distinctive quality of indicating the position and equation of a term. The golden ratio found in the Fibonacci sequence is found in many natural forms, including DNA base and gene sequences, spiral galaxies, renowned art masterpieces (such as Leonardo da Vinci's *The Last Supper* and Salvador Dali's *The Sacrament of the Last Supper*), and by classical composers such as Mozart's Piano Sonata No. 1 in C Major. The sneezewort, a wildflower native to Europe and Asia, particularly displays the Fibonacci numbers in the growing points that it has. Logarithmic analysis of the Fibonacci sequence unveils patterns echoing universal behavior. It is a distant clue as to why gematria works, even through the intermediary of alphanumerical transformation, because the human mind cannot escape what sneezewort cannot escape.

Fibonacci sequences do not necessarily start from zero, as I demonstrate using 108 Hz in chapter 17, but emphasize a *harmonious* system of numerical ordering that imbues any living organism.

Arcturus and the Angelic Realm

Arcturus, meaning "Guardian of the Bear," is one of the brightest red stars that can be seen from Earth. With a mass larger than our sun, Arcturus radiates 113 times more brightness in the infrared and other frequencies of the electromagnetic spectrum compared to the sun.[2] One study proposed that Arcturus, together with some of the brightest stars in the night sky, may have formed beyond the confines of the galaxy, suggesting that it could be up to 12 billion years old.[3] In *Star Names and Their Meanings*, Richard Hinckley Allen explores several mythological references to Arcturus.[4] Its classical name in antiquity among sky watchers translated as the "Keeper of Heaven," the "Lance-Bearer," the "Watcher," or the "Guardian."

In the 1933 Century of Progress Exposition in Chicago, the organizers wanted to open the show with extravagance. At 9:15 PM on May 27, 1933, light was captured from Arcturus using four telescopes placed in different observatories and focused into photoelectric cells. The energy from the photoelectric cells in turn worked to turn on the main spotlights to open the show!

In the beginning of 1928, psychic and medical clairvoyant Edgar Cayce mentioned Arcturus in over 30 psychic readings as a "stargate" to higher realms of consciousness.[5] The following four excerpts are sourced from the circulating files of his readings geared to support the research aspect of Cayce's work:

> We find the Sun and Arcturus, the greater Sun, giving of the strength in mental and spiritual elements toward developing of soul and of the attributes toward the better forces in earth's spheres. (Reading 137-4/Case 137)

Arcturus comes in this entity's chart, or as a central force from which the entity came again in the earth-material sojourns. For, this is the way, the door out of this system. Yet purposefully did the entity return in this experience. (Reading 2454-3/Case 2454)

In the following comparison then we find here that with the variations in the Entity's experience using that of the experience in Mercurian forces to Arcturus brings as an innate force in the present experience that of not only high mental force, yet as measured with Arcturian experience, an excellency above ordinary . . . (Reading 105-2/Case 105)

Arcturus is that which may be called the center of this universe, through which individuals pass and at which period there comes the choice of the individual as to whether it is to return to complete there—that is, in this planetary system, our sun, the earth sun and its planetary system—or to pass on to others. This was an unusual step, and yet a usual one. (Reading 5749-14)

More than just an ancient and powerful star, the reemergence of Arcturus in literature, media, astronomy, and mythological and cultural references indicates an active connection to humanity. In a spiritual context, I intuit the frequencies of the nonphysical forms of Arcturus manifest as angels, guides, and unseen inspiration that can have a profound effect on people's lives. The Arcturus realm is seen as a paramount principle of human evolution because the angelic frequency helps us develop the kind of love and hope we seek in the world. Following this line of thought, the Arcturian epiphany reveals that our existence is intertwined; one cannot exist without the other. We are interdependent on cosmic and atomic scales, and the structure of the physical entity is built by cosmological substances. Our 12-stranded DNA is a cosmic archetype of this principle.

Setting out to visit certain sections of the Sacred Navajo land in Arizona, I came across a special site with a tour group. Taking a photograph of the heavens above, I saw a face flash in my mind as I took the picture. It

wasn't until two years later, when I was randomly inspired to revisit that specific picture, that I recognized the face within the image. The print is located on plate 14. The photograph shows an extravagant view of an orb displaying the shape of a head with eyes and mouth indentations nested in photonic symmetry. Although I cannot determine the source of this entity, I intuit that this hologram represents a type of angelic frequency that manifested at the moment I captured the photograph. Why would this non-physical being appear in the picture? Opinions will vary, but I believe this being, that is interconnected with who we already are, wanted *to be seen and known. This particular entry into the physical realm serves as one form of expression of the Arcturian language. Moreover, it provides evidence of the multilevel Universe containing cosmological levels that inform the physical plane.*

Morphogenesis

Morphogenesis is the motion, signaling, and interaction among all separate parts to become one totality. It is differentiation within evolution. Here, the culmination of the salient teachings of Plato, the sacred geometry cornerstone of the golden ratio and Fibonacci sequence, and our intrinsic goodness and deep empathy by which we, connecting with the angelic realm, comprise the whole and the context in which our DNA and our being can naturally evolve. As a result of the continuous interaction between genes and environments, human DNA evolves, leading to the emergence of new variations.

When I think about morphogenesis in the context of life, I am reminded of minions--small, yellow child-like characters from the *Despicable Me* franchise, known for their unique unintelligible language. Their language reminds me of communication between one cell and another that causes a morphogenetic vibration to DNA and the rest of the body. Like the minions, DNA self-assembles and organizes into different clusters and shapes and is easily affected by emotion. If one cell is feeling sad, it is joined by a group of cells who come to salve and save the day. Angelic frequency is felt as a childlike spirit of the minions that all share a common determined goal of sustaining a unified field as part of the plan of the *Uni*-verse, or "one song."

Tetragrammaton

The Tetragrammaton refers to the Hebrew four-lettered name Yod Heh Vau Heh, transliterated as YHVH, first appearing in the Book of Genesis 2:4 as the deity Lord God, Jehovah Elohim. The background of how the Tetragrammaton came to be stems from the complexity of the naming convention in Judaism. It was considered profanity to pronounce the holy ineffable name of God. Therefore, alternative names for the Tetragrammaton were used when reciting scriptures and may have resulted in the loss of the original pronunciation of the Tetragrammaton. The "divided name" of God is called the Shemhamphorash. The Shemhamphorash comes from the book of Exodus (14:19-21). Seventy-two angelic names are taken from triads of the Hebrew characters in these verses. When all 72 angel names are merged, the Shemhamphorash is commonly written as a spiral.

The story of the Tetragrammaton spindles aspects of symmetry, angelology, divinity, geometry, esotericism, and the implicit power and meaning of words. As a quintessential model, the history of the Tetragrammaton illuminates the intention, logic, and reasoning grounded in the numerical analysis of an alphabet. The fact that Tetragrammaton shares an isopsephy with Plato, Fibonacci, Arcturus and morphogenesis is dumbfounding to me. Through the application of gematria, a collective pattern emerges, revealing that the multilevel Universe is comprised not only of physical matter, intellect, psychology, and cosmological entities, but also of a deeper interconnectedness. Most importantly, it reveals that these elements function together indispensably as a living, breathing organism.

The following meditation is geared to detox the senses and prime your genetic field to tune into the angelic language. It will activate a portal according to your belief system so that you can progress in your continual 12-stranded DNA cellular evolution.

Angelic Frequency Water Meditation

Visualize your body as the ocean or a body of water. With each inhale, imagine generating a wave that moves from the ocean toward you. As the wave starts to rise, envision 12 intertwined helices and allow yourself to feel the answer to the question that follows in the prompt. It is not necessary to accurately depict what a 12-stranded DNA looks like, although with time, you will come to know and be informed. If it is helpful, you can visualize the dodecahedron in plate 4. Try not to overthink it, allowing yourself to feel the first emotion that comes up from the question without judgment. After a couple of seconds, exhale and allow the wave to go back into the ocean.

> Inhale. Visualize the 12-stranded DNA. What does **peace** feel like? Exhale, and let the wave go.
>
> Inhale. Visualize the 12-stranded DNA. What does **light** feel like? Exhale and let the wave go.
>
> Inhale. Visualize the 12-stranded DNA. What does **kindness** feel like? Exhale and let the wave go.
>
> Inhale. Visualize the 12-stranded DNA. What does **forgiveness** feel like? Exhale and let the wave go.
>
> Inhale. Visualize the 12-stranded DNA. What does **ecstasy** feel like? Exhale and let the wave go.
>
> Inhale. Visualize the 12-stranded DNA. What does **compassion** feel like? Exhale and let the wave go.
>
> Inhale. Visualize the 12-stranded DNA. What does **radiance** feel like? Exhale and let the wave go.
>
> Inhale. Visualize the 12-stranded DNA. What does **sacrifice** feel like? Exhale and let the wave go.
>
> Inhale. Visualize the 12-stranded DNA. What does **purity** feel like? Exhale and let the wave go.
>
> Inhale. Visualize the 12-stranded DNA. What does **death** feel like? Exhale and let the wave go.
>
> Inhale. Visualize the 12-stranded DNA. What does **rebirth** feel like? Exhale and let the wave go.
>
> Inhale. Visualize the 12-stranded DNA. What does **love** feel like? Exhale and let the wave go.

Allow the body of water to become calm once again. The frequency in your DNA is infused with the feelings that have come up for you. Your body is attuned to the appropriate frequency and is able to receive information from subtle angelic vibrations.

Angelic Ferocity

Dodecahedral DNA and the
Language of Angels

DNA, in its natural or modified state, holds significant potential in the realm of nanotechnology due to its innovative properties, including self-assembly, sequence programmability, and structural stability. An astonishing physical property of DNA is its electric conductivity, which has been further elucidated by advancements in understanding its capability to absorb electromagnetic waves. One paper annotated how an elementary physical particle absorbs electromagnetic waves within its structure, stores the information, and then subsequently emits it.[1] This mechanism reveals two noteworthy concepts: the interconvertible nature of biological life that evolved from DNA and the outflow of information derives from cosmological energy. This multilevel cosmological entity follows its own parallel laws and implicit logic, which humans classify as science.

Although the unification of the psychological, physical, biological, and cosmological realms comprises the unfolding Universe, it is essential to keep the distinction between cosmological and physical life or matter separate in order to observe the nature of a greater mystery. In the electrical conductivity example, a variation of this cosmological mystery is evident in the interchangeable design of electric and magnetic fields and the absorption and emission processes. Significantly, the process of emission and absorption itself is a mysterious phenomenon. If we can accept that the cosmological substance possesses imperceptible parameters, we might uncover a conveyance through which the lower world (DNA) can receive from the upper world, the cosmological world. In previous chapters, I referred to this transmission as angelic frequency.

I'm no different from you. We come from the same ancestral roots and celestial connections. Recognizing that we are moving through a critical time in human manifestation, we have the opportunity to model genetic evolution based upon angelic teachings. Our light mirrors their light. Each human has the potential to create miracles that benefit themselves, angels, and humanity.

It's important to understand that angels manifest as a frequency to humans according to the belief system of the person with whom they are interacting, picking different mediums like sound, animals, numbers, dreams, nature, and visual images. As we learn to embrace these daily engagements as communications, our beliefs influence not only our reality and creative capacity but also the evolutionary development of human DNA itself. The disposition of angelic vibration attunes human DNA accordingly. Call it aetheric transfer or entrainment; cultivating trust in intelligence becomes an integral component of our somatic and psychospiritual makeup.

My relationship with angels began and continues through intuited presence, visions, animal messages, sounds, the structure of water, numbers (especially alphanumeric ciphers), divine messages, and messages in dreams. The language of angels is distinct to me, though subtle, like the angle at which a bird's wing cuts into the forward flow of air. When I laugh, cry, experience wonder, discern patterns, or am imbued with courage and strength, I am in communion with angels. I understand our connection and bond as a manifestation of love. Sometimes their care comes as a clear knowing in my memory. Other times, I take action with their wind in my sails. Most often, I simply listen to birdcall-like information. While I'm familiar with formal angelology, my purpose in this lifetime isn't to arbitrate the multiple human interpretations of angels. I serve as an active medium for human-angel interaction, furthering the evolution of angelic frequency on this planet.

John Friedlander's *Recentering Seth* depicts the nature of angels better than Caravaggio, Michelangelo, or any official angelology: "Angels have the same consciousness as humans but do not suffer. . . . They just naturally flow toward whatever their next step is without any obstructions. They help create all physical forms and events . . . but without a self-reflective ego. Their awareness never was and never will be dualistic the way a human's is."[2] This sentiment highlights the interconnectedness between angels and humans, with both realms interwoven in a collaborative consciousness. As entities transcending planetary boundaries, angels' intelligence exist as a frequency, even though they might be considered extraterrestrials on Earth's physical plane.

I don't know whether angels have DNA or not, but their existence is part of the same double-helical mode of transmission, suggesting a possible aetheric DNA. Angels are akin to elementals with fluid shapes, capable of existing in various forms on different planes. In the transition from the astral to the crystal plane, they can assume different guises, including lifesaving instincts, animal spirits or totems, and even subtle messages in our environment, such as a vitalizing song lyric in the background of a store or in your car. Angelic frequency can also manifest as intuition, an unexpected inclination of joy, faith, belief, or self-sacrifice for a greater purpose. In human terms, an angel is a psychoid, a real energy that becomes tangible by our conscious experience of it. In *The Night Sky*, Richard Grossinger described psychoids by proposing, "A psychoid's validity is established solely by our experience of it. If, as we approach its phantom (like a fairy or animal spirit or even a neutrino), it continues to grow more real and autonomous, then it *is* 'real,' or exogenous, and the universe itself is leading us to it."[3]

Humans have a natural ability to interface with psychoids, or angelic frequencies, through multiple channels such as animals, water, sound, dreams, and numbers. My personal journey began as a woman born in Kyiv, a troubled power spot, and educated as a biochemist who pursued a career in the biotechnology industry. Then I awoke to the fact that I was handling angelic, aetheric material. This revelation led me to question the practice of materialism and seek the truth that lay before me. It was a life-changing revelation and turning point. If I would have continued to stay on my former trajectory, one might have encountered a bio-formulation developed by me rather than a higher-frequency prayer. Both would have been DNA, but what I want to share is the vibration of *free* DNA, not some captured patented configuration of it like a lion in a zoo.

Considering DNA as a form of evolved consciousness, it isn't limited to genetic instructions, pattern recognitions, and sequences alone. DNA can absorb angelic information due to our cerebral and genetic memory and the unconscious realm. This connection extends to the aetheric frequency above the material world, and ultimately to the aetheric spirals at the source of DNA messaging.

How do I know this? Although my scientific background does not include angels in its repertoire, it acknowledges that noncoding memory is passed down generationally and can be shared telepathically. We may train our minds to be receptive to ethereal frequencies by meditation, prayer, chanting, energy healing, connecting with the soil, and participating in the ceremony of life. As Rupert Sheldrake noted, more than 80 percent of people have reported experiencing seemingly telepathic moments, such as thinking of someone for no apparent reason who then calls, or knowing who is calling before picking up the phone.[4] Angelic frequency and communication is no different from any other telepathic or intuitive exchange.

Angels have the ability to become human if they choose to do so, embodying angelic beings who partake in human experiences. This choice allows for the emergence of a new form of humankind—a new form of being, a hybrid of human and angel. These hybrids inherently gravitate toward goodness.

Many associate angelic frequency with high vibration of ascension, but it also embodies the cohesiveness known as love. Simply put, it's a shift in core and system intelligence. To understand this form of love intelligence, we must rewire *everything* that we know about both love and intelligence, beginning with the way we see ourselves, the words that we use, how we treat others, and the focus of our attention and breath. Once the mind acknowledges the existence of angelic energy, it can become a sensory experience and be transformed into personal angelic ferocity.

ANGELIC FEROCITY AND MY ANGELS

Angelic ferocity is sheer magnificence with a kick: an energy that strives for a higher level of complex intelligence, vibrant goodness, natural self-lessness, and a touch of innocent unpredictability. It embodies the essence of being a spiritual ninja—navigating and influencing energy fields in our interactions, both in time and space. Angelic ferocity involves pursuing projects with sincerity and the intent to make a positive contribution to society, fostering energy flow in day-to-day activities, and cultivating a natural magnetism that inspires joy, optimism, and nourishing vibrations without attachment to dualistic outcomes. This self-tendency seamlessly integrates into all of life, actively building bridges for everyone's benefit. The key to embodying angelic ferocity lies in performing these actions instinctively and unselfconsciously—a spiritual circadian rhythm. To explore your own angelic ferocity, consider these questions: In the absence of money and power, what would your actions be? Where would you focus your attention? What spiritual approach would you adopt?

My relationship with angels began at the moment of my conception. My birth was a miracle in more ways than one; I arrived early and was initially assumed to have been another miscarriage caused by the radio-active fallout—DNA jumble—from the Chernobyl accident in 1986. Chernobyl is not far from Kyiv, nearly sixty miles. For my mother, my birth was a matter of life or death. During my birth, her heart stopped beating for two minutes, and she was thankfully resuscitated back to life.

I had a guardian angel from the time I was six months old, when I was learning how to crawl. One day, I accidentally stuck my thumb into an electrical outlet under my parents' bed and was electrocuted.

When my father found me, I was surrounded by smoke. The emergency technician who arrived with the ambulance couldn't believe I was still alive. Even now, I carry the scar on my thumb as a reminder of that miraculous experience.

My father, Michael, resonated with the leadership archetype of angelic behavior, similar to Archangel Michael, "One Who is Like God." As I grew and learned to walk, my insatiable curiosity led me to embark on adventures that were both amusing and dangerous. On one occasion, I disappeared from our apartment and managed to open the door, take the elevator, and venture outside. My dad spotted me from the balcony, proceeding into a bustling speedway in the Obolon neighborhood. He hastily rushed downstairs and scooped me up just moments before I was about to step into the first lane of oncoming traffic.

When I was five, my grandmother Anna exuded the first exogenous angelic consciousness I can remember. As I played on the grassy lawn outside my parents' apartment in Virginia after we moved to the United States, I noticed a man with unkempt dark brown hair, a tan buttoned-up shirt and ragged shorts sprinting *at* me. I froze in fear, sensing his intentions to abduct me. My grandmother, who was close by, saw him too. Despite being unable to run or speak English, she hurriedly walked and stood in front of me, waving her finger at him with an authoritative and clear, "No." Her aura was white and shielding. He came to a halt, nervously snorting and pointing to his ears with a distorted smile. "Be careful, I'll grab her ears," he warned. In that moment, angelic intelligence communicated a telepathic awareness, fending off danger.

At the age of eleven, my father enrolled me in a tennis summer camp as a punishment for secretly teaching myself how to hack software programs. Tennis quickly became a passion and a decade-long competitive regime. It was the perfect way to "hack" the real world, blending physics, quantum motion, and social interaction. I became enthralled by the control and manipulation of spin, power, friction, and carbon fiber—all centered around a yellow fuzzy nucleus, the ball. Spending hours rallying the ball felt like a psychic connection, providing a thrilling sense of focus and thought control. Immersing my mind in the ball was comparable to a spiritual practitioner exploring a spiritual visualization.

But as time went on, my need to win became an addiction. I played for the tennis club in college, then coached my high-school tennis team until a subtle snapping sensation in my lower back during a serve left me with a debilitating injury. For nearly a year, the pain was severe enough to make me doubt if I would ever walk normally again. Unable to continue playing, I redirected my attention to science and research throughout college. In hindsight, this shift felt like a divine intervention—a way for my angels to help me heal from a toxic relationship with tennis and guide me toward becoming a trained scientist in 2010.

Following my dad's death, my grandmother, Roza, who was already battling cancer, experienced a rapid progression of her illness, exacerbated by her overwhelming grief. We were both displaced, but I dedicated myself to helping her transition to the next life. This marked the beginning of a conscious journey in communicating with our ancestral angels to heal her soul. During the last six years of her life, our daily conversations deepened our connection, strengthening our psychic bond. We became spiritually one, transmuting our shared grief into an acknowledgment of death and *anticipation* of rebirth. Over time, Roza shifted from a mindset of having "nothing to live for," to finding purpose and value in life. At the end of her life, she said, "I'm happy. I got to live another day. I'm grateful for everything that God gave to me in my life, and I forgive all." Born in 1936, Roza had endured food scarcity, witnessed mass genocide, and recovered from an anti-Semitic regime.

ANGELS SPEAK

It is an illusion that each success, promotion, and achievement is accomplished by individualism or self-proclaimed enthusiasm. Angels are the silent partners supporting each step in our lives, whether they get recognized or not. What people understand as a "miracle" or "synchronicity" is communication of the angelic frequency.

The language of angels is characterized by directional vibrations in meaning, timing, and events. Seemingly random occurrences, such as a fluttering butterfly on your walk, a license-plate number that catches your eye, a logo on a shirt, or a message on a billboard are all elements of nonrandom frequency that are waiting to be acknowledged. By associating your consciousness with the frequencies explored in this book, you connect with your angelic frequency.

Genetic sequencing—the technology of reading DNA—is one of the most advanced technologies in life sciences. DNA sequence-pattern similarities reveal homologies proving we all share a common evolutionary ancestor. Yet, sequencing technology is not only used to read the structure and function similarities in DNA or to identify genetic mutations. It can also explore the pattern sequences of angels. An incomplete sequence is not random or unordered, a concept applicable in biochemistry, theology, and spirituality. In DNA applications, a string sequence is converted into a numerical value to form a matrix model. The encoded model is used for tracking algorithms within neural networks. In this context, pattern recognition and its discrimination of sequences can be viewed as a discrete, but imperceptible language. In other words, we can access mood indigo, or a unique and enigmatic attention and connection between the physical and spiritual planes by the spaces in between. It's similar to connecting a jumble of dots and numbers on a map to take a look at the image, except the space between the content is what *makes* the content visible.

Through the subtle activity of my magnetic field, I have discerned the following angelic languages that evoke emotional and physiological shifts within me. These experiences may typically be viewed simply as tragedy, humor, courage, or feeling stuck. Over time, I have observed and organized these patterns as a spiritual encoded matrix model, guided by angelic companions. In each language, I have witnessed hardship become a source of transformation and closure—a shift as dynamic as singing away a stormy sky to reveal a serene, clear one, something I have experienced through my harp playing.

Language of Regression

Regression is characterized by loop thinking, which leads nowhere. Freud described regression as the temporary reversion of the ego to an earlier state of development. It's a swampland where stagnant water saturates the land. Have you ever found yourself caught up in thoughts of how you could have acted differently in a particular situation, only to find that your energy becomes stagnant? Or experienced cyclic thinking that intensifies your feelings, leaving you "dragged down" by them? Our spirit urges our mind to break free from regression because it, too, feels trapped.

In such situations, a person generally faces two options: One could either maintain the same course of action or shift oneself. It doesn't have to be a dramatic shift, it can be as simple as altering the wording in an email or a casual conversation. While acknowledging that "everything counts" may not be an angelic message, it is a response to the speech of angels. Those actively pursuing personal development organically strive to invoke angelic energy in some way. Although this connection may not always be apparent to others, that's between the person and the angels, not you or me.

Angels operate as spiritual guides, helping steer individuals along paths filled with unexpected twists and turns. While these paths may not initially make sense, their purpose becomes clear with time. What humans perceive as "right" can be at odds with the celestial intent. A missed flight might prompt a conversation at the airport that results in two people falling in love. That's angel talk. Getting fired or laid off from an employer sets you on the path to a better opportunity. The sudden loss of a loved one is followed by a new relationship. Angelic intelligence is the unseen inspiration that accentuates we are where we are meant to be and encourages us to soar in life. By committing to positive-intentioned decisions, we act on the guidance of angels—the whispers, prayers, and trust—and embrace whatever may come. In her review of a book by Simone de Beauvoir in the *New York Times* in 1966, Irish novelist Brigid Brophy wrote tellingly, "Mlle. De Beauvoir has often (though not as often as she might, had she exercised her intellect and imagination more ardently) been on the side of the angels; but the method by which she arrived there must often have been enough to make the angels weep."[5]

Ruslana, too!

I slept in the guest room of my own house on November 25, 2016, packed a suitcase of my father's belongings and three days of clothes, and left my seven-year-long marriage the next morning. It's difficult to narrow down the reason for my aspirational thrust out the door. It was a mixture of things: a growing sense of hopelessness from an apocalyptic cult-like church, feeling like a Cinderella sort of wallflower in my marriage, or the last goodnight wish from my ex-husband, "Don't be so needy," as he faced away from me when I desperately sought connection and emotional support while mourning my father.

My decision to leave was as bright as a light bulb illuminating a long-darkened room. Although there were valid reasons to leave earlier, I lacked the strength to do so until that night. A fairylike chorus serenaded me into the guest room. The angels showed me everything I needed to know in my dreams that night. With newfound clarity and confidence, the angels saved both of our lives by breaking the cycle of stagnation. Leaving was an act of service—neither of us was growing together, and I had to summon the courage to embrace change. I longed to redefine myself beyond the expectations or notions that did not resonate with my true self. I refused to let my life be defined by the belief that, as a Ukrainian woman, I was obligated to marry within my family's ethnicity or conform to a crusty old Soviet mindset that marriage and children were simply "what we do." I sought to trust my own ability to designate sacredness in life without the constraints of a cult-like church. During those seven years, I discovered the existence of a personal hell—the reality I had created for myself by feeling trapped in a destructive cycle. It was the guidance of my angels that ultimately nudged me to move on and break free from these limiting expectations.

I wanted to share this story and encourage readers to believe in their instincts—being a form of memory, passionate creativity, and ageless knowledge. I transitioned from a place of perceived helplessness and neediness to a newfound sense of inner strength and hope. My neediness was a cover for my untapped power, a fear to evolve.

Our behaviors develop within social experiences and ecological context. Sea turtles instinctively know to find their way home to the ocean upon hatching. Birds know to migrate south for the winter, and collies naturally know know how to herd sheep. Learning is an inherent instinct. Embracing change and actively pursuing growth, while upholding our core values, leads to fullfillment and contentment. Angels act as a voice guiding us through regression thresholds because they exist without self-reflective egos. To begin a journey of working with regression, consider being a surrogate angel by engaging in acts of service or helping others. An exalted act of selflessness and care can be the off-ramp from mental and emotional gridlock.

Language of Interventions

While I run "mere" marathons, my partner is an ultramarathoner. On a forty-mile run on the Capital Trail that links Richmond to Williamsburg, Virginia, I met him and his running team to provide nutritional support. Shortly after resuming their run after a snack break, my car battery died, and I realized I didn't have jumper cables. Despite the predicament, I remained calm, intuitively believing that someone would come to my aid. Almost miraculously, three separate people in the park approached me simultaneously, offering assistance. My angels were showing me that help is always available.

Interventions serve as conduits to developing trust in angels. We can't predict when they'll occur—sometimes they may result in heartbreak, while other times they feel like winning a lottery. Despite the unexpected change, try to stay grounded in the present moment and practice this mantra: "I am safe. I trust life and my angels. I flow with these energies." Take a few deep breaths, inhaling and exhaling slowly.

Language of Laughter

Laughter doesn't necessarily originate from a comedic source. It can also arise from truth, irony, or angle of perception. Laughter serves as a reminder to relax and detach from the intensities that we create for ourselves. It can be a release valve when we feel like crying or getting angry. While what we value and do is important, our existence is much more than the output we can offer on any given day.

A good laugh is particularly beneficial when we find ourselves feeling pressure or stress from focusing intensely on one task, helping us avoid becoming overly neurotic or compulsive. Take a moment to recount your day's events. Write them down and say them out loud. For example, "Today, I woke up and talked with my cards. I sat in my hot tub and meditated. I made coffee, jumped around a few times for fun, and wrote for ten hours. After about ten hours of hard work, I wanted to cry. My friend came over to give me a hug, and I felt better." When I repeat the story out loud, I smile because it has a very different effect when I *hear* what I did versus experiencing what I did. Reciting your day's story aloud may evoke a smile as you perceive it from a different angle.

Just as breathing creates space between thoughts, laughter offers coming relief in the ongoing narrative of our lives. It allows us to separate ourselves from our experiences and make adjustments. Angels, too, share in the healing power of laughter.

Language of Tears

Tears have meaning and are an essential aspect of emotional expression and the soul's connection with its surroundings. My former Ayurveda teacher Katie Silcox once said, "Emotions are movement of the soul. When we feel mad, sad, happy, or any defined emotion, the movement of energy represents a healthy body. The word 'emotion' itself can be broken down to 'e' to represent energy and 'motion.'" Tears are a healthy part of the self's feeling oneness with its surroundings. They are energy, too.

There is a significant difference between crying because things are not going according to our ego's preferences and engaging with our surroundings through the ego. I can't help but cry for the beauty of the natural world. However, when things don't go "my" way, it doesn't bother me as it once did. I trust my guides and angels.

Whether we feel moved by fulfilling connections or experience emotional hardship, we can awaken the angelic frequencies through our words and thoughts, even amid tears. Tears are a sacred soul distillation. They prepare the body for the soul's instructions, similar to how incense, silence, dance, meditation, personal journeys, or music prepare us to express, reinforce, and transmit our values in ceremony. Above all else, tears show care in what we deeply value.

I recall a time when I experienced emotional anguish after an unexpected and crippling phone conversation from a family member. Feeling a heated tear rolling down my cheek, "this is beautiful," I whispered to myself, feeling the truth behind this perturbing conversation. Despite the circumstances, I trusted we would both heal in our authentic ways and transform our relationship into a deeper bond. Another moment where I experienced conflicting emotions but responded with spiritual awareness was with my employer. While enduring a daunting end-of-year performance evaluation, I felt the perplexity and dismay in response to my supervisor's negative feedback without acknowledging one of the hardest working periods of my life. Feeling the heat rise from my chest to my swelling face, I responded, "I am grateful," shifting my mindset towards soul expansion. A separate occasion when I experienced physical adversity was when my partner and I hiked the Hundred-Mile Wilderness Trail in Maine. When I cried from emotional despair due to hunger and cold, I said to Gaia, "I trust you." These statements of acknowledgment activate my mind and heart to vibrate with the cosmic chord. It takes courage to say it and believe it, but we change the course of our tears and events in our lives through such thought forms and beliefs.

Closing the circle of each scenario, my family's fallout reconciled as a deeper bond; I left my employer to write this book; an angel in the form of a woman appeared in the woods with a table of food and saved our day with sandwiches, Doritos, Sprite, Oreos, gummy bears, grapes, applesauce, and a Bud Lite. My partner and I had been sharing seven hundred calories of food between us for the two days since we ran out of supplies. We were easily exerting at least ten thousand worth of calories per day for rucking forty-pound backpacks on one of the most generally favored and technical sections of the Appalachian Trail. But that is a story for another book. I stared at her while she told us she was planning on leaving four days earlier but had a hunch to stay and feed the hikers. I wanted to kiss that angel.

Acknowledge each tear. Even if silent, treat it as a gateway to bounty and part of your evolutionary journey.

Language of Courage

Someone once told me, "Game recognizes game" in response to my appreciation for the hard work that he does for the community. I said, "I don't see it as a game, competition, or comparison. Love recognizes love." I noticed the tension dissipating from his shoulders, eye sockets softening and his pupils constricting. I pay attention to eye activity during my interactions with people because it can reveal valuable insights into their underlying emotional state and inner thoughts. It's not so much a matter of pupil dilation (which reveals tension or worry) or constriction (relaxation), but similar to numbers, eyes tell the story from within. He thanked me for cuing him and shook off the momentary addiction of winning. Winning is a concept that humans have invented to accentuate skill, technique, and knowledge and to express the old mammalian concept of territory. Courage, however, transcends these, as it is rooted in sacrifice and cosmic strength.

My first food truck event was the very next Sunday morning after my last day at the lab, at a small local festival. It was the same day the generator that came with the truck died after only two hours. "A quirky beginning," I thought. Rather than feeling defeated, I felt an energizing rebelliousness and determination to handle my business with humor, eccentric vibrancy, and a "get it done" attitude. The next day, I went shopping for new generators and selected the best one on the market. The following week, my truck had a brand-new paint job, leaving the 1988 relic gleaming and probably living her best years.

I led my entrepreneurial experiences with my full heart. Although I wasn't experienced in business or the food industry, I worked my way up just by being honest and sincere. I became more comfortable with customer relations, event coordinating, and marketing.

I quickly established a strong network of fellow entrepreneurs by meeting them through the events where I was vending. Although my first food truck season was short-lived because of the approaching winter, I learned the ins and outs of business within six months, and I was a one-woman show. I didn't know what the next season was going to bring, but I felt a cosmic bang coming.

After a trip abroad, and the same week as my birthday, a gift from the heavens fell right onto my lap. The timing of it had to be as precise as a successful jump over a canyon. It was an opportunity to move into a brick-and-mortar location, my own shop.

That same season, I was interviewed as a contestant for Virginia's best food truck by the Food Network, which chooses contestants from the top fifty in the country. After I was selected as a candidate, a stream of other local acknowledgments and awards paraded in. My original dream and vision for community in the form of my coffee shop and food truck were bright and sparkling, the small but sweet dazzling stars in the night sky.

During my first visit to Colorado, I met a medicine woman while camping with a group of woman celebrating Eros in ceremony. She and I talked about the soil, the dead and the living, and the womb. She invited me to participate in a ritual that would challenge my beliefs about courage. I dug a hole that was large enough to submerge and support my body. I slept in it overnight. I felt the ground supporting me. I felt myself as part of the organisms, dead and alive, and the sound of the underworld.

As dawn broke, I lay in silence, enveloped in fog, listening to the ground and the subtle movements in my body. Beginning with the womb, I felt a ring of "chargeness" send *information* to my heart and my feet at the same time. Unlike the charged feeling of a spike of sudden energy when running, drinking caffeine, or even the thought of fear or worry, this profound and gentle feeling was like a single drop of water forming ringlets in a lake encircling the rest of my body in the sacral. My mental awareness of it initiated a pleasant ringing cerebrally. This awakening felt like angels singing around my aura, levitating yet connected to the ground. In that moment, my former self became compost, and the new part of me was *in* the soil. I could not only feel myself but also the organisms that supported the soil in that moment. Reluctantly leaving the embrace of the earth, I left my sacred offering to Gaia that morning, my copper bracelet, as a token of my loyalty and thankfulness for accepting and activating my DNA along with all of the other bacteria and organisms that depend on its soil ecology for prosperity. Humans too.

That experience awakened a new realm of courage in me. I didn't see myself as a competitor anymore. I saw myself as a temporary inhabitant devoted to a mission.

PART 3

ADVANCED
PRACTICE
Back to the Drawing Table

Introduction to Part 3

When Pythagoras passed away, he didn't leave any writings behind. His legacy was to instill his students with knowledge about the foundational harmony between numbers and the universe. After all, the first civilizations recognized the way letters can be associated with numbers. Western philosophy originated in ancient Greece, and its key figures Socrates, Plato, and Aristotle were Pythagoreans who believed that everything in the Universe can be interpreted through numbers, and consequently the letters that make up the word will correspond in numbers.

Eleftherios introduced and inspired me to use the lexarithmic* theory as one of the targeting probabilistic methods for success in my future experiments. I've approached this technique in passing conversations, bridging time to historical events, concepts, or people, and even anatomizing a name to gain a deeper comprehension and insight into the object. For this book, this concept offers new insight and perspective about the 12-stranded DNA. It's not a conclusion, but a means to analyze further.

Paul Valéry would have agreed with me when I say the work that interests me deeply is the work that incites me to picture the living and thinking system that produced it.[1]

---

*Lexarithmos is a term that refers to the relationship of letters and words with numbers. Lexis (λεξις) means "word" and arithmos (Αριθμος) means "number."

Gematria Meditations

In this section, I share 32 variations of gematria meditations that demonstrate aspects of unique parity, nondualism, and an inventive approach to observing hidden patterns in nature. Please keep in mind, there is nothing that you literally *do* with the numerical values. Moreover, the initiative emphasizes a meaningful and eclectic approach to examine numerical equivalence amid diverse ideas and concepts. The method illuminates the symbolic all-encompassing nature of the dodecahedron and creates possibilities for an innovative way to observe a thought (in the form of a word or phrase) objectively using symmetry and numbers. After all, there are 520 ways to prove the Pythagorean Theorem.[1] This alone should encourage you of the implicit exuberant possibilities that an individual word contains more than one link to another word that inherently has the same gematria value.

While the phrases share diversity in meaning, they are also unified by theoretical concepts, cosmological themes, and imperceptible parameters oriented towards the 12-stranded DNA. This aspect demonstrates the non-dualistic nature of the dodecahedron discussed in chapter 4. The phrases do not relate to a classical text, scripture, or name of a deity. By gathering scientific, historical, and cultural information from a variety of sources and facilitating a careful analysis of the Greek terminology and evolution of language under the guidance of my mentor Eleftherios while writing this book, I received telepathically knowledge from a source that I cannot trace. By nature, I cannot provide a traditional citation for the knowledge and information that came to me. This practice of this concept is similar to the work of DNA researchers who intuitively assigned musical notes to DNA sequences. When the resulting song is played, the melody not only organically resonates with the emotions of the listener but one can detect a hidden pattern in the song.

HUMANITY'S FUTURE EVOLUTION
Gematria of the 12-Stranded DNA

I explore the gematria of the dodecahedral DNA through a variety of expressions to create multiple opportunities for comparisons and a queue of meanings. To access more knowledge, I had to step out of the technical, constrained limits of using only "DNA" to experiment with phrases such as "dodecahedron deoxyribonucleic acid," or adding an article to the phrase: "*the* dodecahedron deoxyribonucleic acid," or "the dodecahedron DNA," or "dodecahedron DNA activation."

After deciding on the wording of the original phrase, I calculate its gematria value using the Hellenic code: adding up the number values of the word to form a single number, according to the table shown below.

THE HELLENIC ALPHA-NUMERIC SYSTEM

A, α = 1	I, ι = 10	P, ρ = 100
B, β = 2	K, κ = 20	Σ, σ = 200
Γ, γ = 3	Λ, λ = 30	T, τ = 300
Δ, δ = 4	M, μ = 40	Y, υ = 400
E, ε = 5	N, ν = 50	Φ, φ = 500
Ϛ, ϛ = 6	Ξ, ξ = 60	X, χ = 600
Z, ζ = 7	O, o = 70	Ψ, ψ = 700
H, η = 8	Π, π = 80	Ω, ω = 800
Θ, θ = 9	ϟ, ϟ = 90	ϡ, ϡ = 900

I follow the translation of a Hellenic phrase with its Latin version and then its English translation. I looked for a separate phrase corresponding to the original gematria number of the original phrase. How I determined such phrases is complex: it includes my study of the Hellenic logos, that being the intentional use and application of the word as the fundamental substance of the Universe, guidance under the mentorship of Eleftherios, and divine inspiration. There were times when I didn't know why I was being led to certain ideas and concepts, later to experience their corresponding gematria have a match with the original phrase. My angels cued me.

When comparing gematria values, words and phrases that contain numerical equivalence may not be conceptually identical, but they are correlated. In any group of words or phrases indicating the same gematria, the meaning of the phrases or words are related to the numbers as well as a number's theme. Therefore, the following words are not conceptually identical to "the 12-stranded DNA" or "the dodecahedron," or any phrase pertaining to the original phrase, but in some way related to it, in terms of past, present, and future. Several of the phrases here may have metaphorical meaning. Thus, two separate phrases correlate to the same gematria value deciphered by the alphanumeric system of the Hellenic language. The Latinized Greek and English phrases repeat natural gematria of the original Hellenic phrase or word. This relationship is referred to as *isopsephia* (or isopsephy), which presumes an equal value between the phrases.

The Latin phrase is directly translated by each character from the Hellenic phrase. Both Greek and Latin are fundamental roots of the English language. I've included the Latin phrase because over half of English vocabulary is penetrated by either the Greek or Latin language. Both phrases are translated into English by their original meanings in this chapter. The phrases are organized by language in the following manner:

> **A. Hellenic**
> **B. Greek in the Latin alphabet**
> **C. English**

There are endless possibilities using this approach. The numerical intelligence of the alphanumeric system of the Hellenic language excavates the dodecahedral from abstract speculation to tangibility, definition, and imagery.

Entry of Divine Grace into Human, the Solar Cradle of Souls

The first isopsephy corresponds to the phrases "entry of divine grace into human" and "the solar cradle of souls." I illuminate them this way: When a person decides to activate their 12-stranded DNA, they acknowledge their equivalence to the divine, as godly, angelic, and spiritual. "Divine" shares a root with the word *human* in Sumerian etymology. *Hu* means god, and *man* means person. "The solar cradle of souls" is another way of saying "sunlight of souls." Light and solar energy create the electric field of DNA (the cradle), where the soul resides. This field is also coupled with the concept of the angelic frequency or ascension.

A1. δωδεκακλωνικον δεσοξυριβοζοπυρηνικον οξυ = 4200

A2. εισοδος θειας χαριτος εντος ανθρωπου = 4200

A3. η ηλιακη κοιτιδα των ψυχων = 4200

B1. dodecaclonikon desoxyrivozopirinikon oxy

B2. isodos theias charitos entos tou anthropou

B3. i iliaki kitida ton psychon

C1. dodecahedron deoxyribonucleic acid

C2. entry of divine grace into human

C3. the solar cradle of souls

Obvious Activation of the Genetic Code, Action through the Genetic Material

The phrases "obvious activation of the genetic code" and "action through the genetic material" share the same gematria value as the phrase "dodecahedron deoxyribonucleic acid," when written in a slight modification than the previous phrase. One is able to activate their DNA because it's an overt system, which is my interpretation of the word *obvious*. A human being is able to ascend by their 12-stranded genetic material to access new levels of consciousness and mental action.

A4. δωδεκακλωνικον δεσοξυριβοπυρηνικον οξυ = 4123

A5. ολοφανερη ενεργοποησις του γενετικου κωδικα = 4123

A6. δρασις μεσω του γενετικου υλικου = 4123

B4. dodecaclonicon desoxyribopyrinikon oxy
B5. olophaneri energopoiesis tou genetikou kvdika
B6. drasis meso tou genetikou kodika

C4. dodecahedron deoxyribonucleic acid
C5. obvious activation of the genetic code
C6. action through the genetic material

Human's Exclusive Genetic Mutation, the Benefits of Silent Meditation

The meaning of mutation is variable in English. My interpretation of mutation in this context means to alter one's thinking. It entails the acceptance of the 12-stranded DNA as our natural design, while adapting to a lifestyle and mindset that supports genetic progression. Because this type of evolution is mental, practices such as meditation and breathwork will benefit your activation.

A7. δωδεκακλωνικον δεσοξυριβονουκλεϊνικον οξυ = 4120
A8. η αποκλειστικη γενετικη μεταλλαξις του ανθρωπου = 4120
A9. τα οφελη του σιωπηλου διαλογισμου = 4120

B7. dodecaclonicon desoxyrivonucleinikon oxy
B8. i apoklistiki genetiki metallaxis tou anthropou
B9. ta opheli tou siopilou dialogismou

C7. dodecahedron deoxyribonucleic acid
C8. human's exclusive genetic mutation
C9. the benefits of silent meditation

The Illuminated by the Holy Spirit, Mental Energy of the Evolution of the Human Intelligence

By adding the appropriate article to the previous phrases that describe the 12-stranded DNA, new values are extracted. The following phrases emphasize that this evolution is correlated to the way the mind displaces thought and the breath throughout the body-mind. By this approach, a person is embodying his or her intelligence to enlighten and "illuminate the Holy Spirit." The 12-stranded DNA emerges as the sine qua non of human evolution, with its activation providing the essential mental energy needed to propel our intelligence forward.

A10. το δωδεκακλωνικον δεσοξυριβονουκλεινικον οξυ = 4490

A11. οι φωτισμενοι εκ του αγιου πνευματος = 4490

A12. νοητικη ενεργεια της εξελιξεως της ανθρωπινης διανοιας = 4490

B10. to dodecaclonicon desoxyrivonucleinikon oxy

B11. oi fotismenoi ek tou aghiou pnevmatos

B12. noitiki energeia tis exelixeos tis anthropinis dianoias

C10. the dodecahedron deoxyribonucleic acid

C11. the illuminated by the holy spirit

C12. mental energy of the evolution of the human intelligence

The Weapon of the Gods, the World of Gods, Receiver of Divine Spirit

The associated phrases corresponding to "the dodecahedron DNA" are "the weapon of the gods," "the world of gods," "receiver of divine spirit." As discussed in chapter 11 on dreamtime, "receiver of divine spirit" holds particular importance. I believe the interpretation of "weapon" in this phrase is equivalent to the meaning of "power." Combined with the phrase relating to the world of gods, the 12-stranded DNA is a human's birthright activated by intuition and sovereignty.

A13. το δωδεκακλωνικον ντι εν ει = το οπλον των θεων = 2684

A14. ο κοσμος των θεων = δεχομενος θειου πνευματος = 2684

B13. to dodecaclonikon nti en ei = to oplon ton theon

B14. o cosmos ton theon = dechomenos theiou pnevmatos

C13. the dodecahedron DNA = the weapon of the gods

C14. the world of gods = receiver of divine spirit

Life of Unknown Origin, Enlightenment

The numerical value of the phrase "dodecahedron DNA activation" shares the same isopsephy as the "life of unknown origin" and "in-lamp enlightenment" (or enlightenment). When the sun sets, all is dark. The only light that you see is emitted by a lamp or candle. This is a metaphor equivalent to the Hermit card of the tarot to describe the mental awakening from the former double-stranded DNA to its archetypal gene.

Accepting one's genetic destiny initiates a journey of resolution and enlightenment. My intuition of the subtle sense of the phrase "Life of unknown origin" could also mean "meant to be partially understood." We strive to understand the bigger picture, but the complete picture eludes us.

A15. ενεργοποιησις δωδεκακλωνικου ντι εν εϊ = 3475

A16. ζωη αγνωστης προελευσης = επιλυχνιος φωτισις = 3475

B15. energopeoisis dvdecaclonikou nti en ei

B16. zoi agnostis proelefsis = epilychnios fotisis

C15. dodecahedron DNA activation

C16. life of unknown origin = (in-lamp) enlightenment

The Master of Nature, Neurolinguistic Programming, the Center of Spiritual Power, Thus Was Human Deceived

Four isopsephies stem from "activation of dodecahedron DNA:" "the master of nature," "the neurolinguistic programming," "the center of spiritual power" and "thus was human deceived."

The last phrase suggests humans were originally designed with the mentality of the 12-stranded DNA, which emphasizes the beginning mark of the most exciting awakening in the collective consciousness of humanity. Activated by neurolinguistic programming, the 12-stranded DNA is a juncture to identity, beliefs, behaviors, purpose, and capabilities—the mode of this book.

A17. η ενεργοποιησις δωδεκακλωνικου ντι εν ει = 3483

A18. ο κυριος της φυσεως = ο νευρογλωσσικος προγραμματισμος = 3483

A19. το κεντρο της πνευματικης εξουσιας = 3483
A20. ουτως επλασθη ο ανθρωπος = 3483

B17. i energopeoisis dodecaclonikou nti en ei
B18. o kyrios tis fiseos = o nevroglossikos programmatismos
B19. to kentro tis pnevmatikis exousias
B20. outos eplasthi o anthropos

C17. the activation of dodecahedron DNA
C18. the master of nature = neuro-linguistic programming
C19. the center of spiritual power
C20. thus was human deceived

The Vital Energy within the Human

The isopsephy of "the vital energy within man" matches the gematria of "activation of the dodecahedron DNA." The journey of this new mentality expands a vitality within. Vitality means fulfillment, meaning, purpose, love, livelihood, discovering new power and wisdom by the sugar-twisted DNA molecule.

A21. ενεργοποιησις του δωδεκακλωνικου ντι εν ει = 4245
A22. η ζωτικη ενεργεια η εντος του ανθρωπου = 4245

B21. energopeoisis tou dodecaclonikou nti en ei
B22. i zotiki energia i entos tou anthropou

C21. activation of the dodecahedron DNA
C22. the vital energy within the human

He Rested the Seventh Day from All His Works

The next gematria value equals "he rested on the seventh day after all the works he created." This phrase depicts the relationship between the divine field and quantum source that animates life.

A23. η ενεργοποιησις του δωδεκακλωνικου δεσοξυριβοζοπυρηνικου οξεος = 6364
A24. ανεπαυθη την ημεραν την εβδομην απο παντων των εργων αυτου = 6364

B23. i energopeoisis tou dodecaclonikou desoxyrivozopyrinikou oxeos

B24. anepafthi tin imeran tin evdomin apo panton ton ergon aftou

C23. the activation of dodecahedron deoxyribonucleic acid

C24. he rested the seventh day from all his works

The Mystery of the Creation of the World from Zero

"The activation of the dodecahedron deoxyribonucleic acid" corresponds to "the mystery of the creation of the world from zero." When we take proactive steps in committing to the inner work—that is, meditating, playing sacred sounds, energy healing—a mystery's closed circle unravels. What was once a mystery becomes palpable and a source of revelation.

A25. το μυστηριο της δημιουργιας του κοσμου εκ του μηδενος = 5594

A26. η ενεργοποιησις δωδεκακλωνικου δεσοξυριβοζοπυρηνικου οξεος = 5594

B25. to mistirio tis dimiourgias tou cosmou ek tou midenos

B26. i energopeoisis dodecaclonikou desoxyrivozopirinikou oxeos

C25. the mystery of the creation of the world from zero

C26. the activation of dodecahedron deoxyribonucleic acid

Psychosomatic Health with Heavenly Vibration

Psychosomatic health, impacted by daily social, psychological, and behavioral aspects, is influenced by our mind and body. By refining one's thought processes to our archetypal genetic structure, we attain harmonic "psychosomatic health," to the degree of "heavenly vibrations" from the Universe and the angelic realm.

A27. η ενεργοποιησις δωδεκακλωνικου δεσοξυριβονουκλεϊνικου οξεος = 5514

A28. η ψυχοσωματικη υγεια με τις ουρανιες δονησεις = 5514

B27. i energopeoisis dodecaclonikou desoxyrivonucleinikou oxeos
B28. i psychosomatiki hygeia me tis ouranies dynamis

C27. activation of dodecahedron deoxyribonucleic acid
C28. psychosomatic health with heavenly vibration

The Outer Personality of the Human Being

The phrase shares the isopsephy with "the outer personality of the human being," connecting the akashic cosmic records within a person and the wisdom that is discovered within.

Conventional science cannot penetrate the akashic field or even know that it exists, but it is a key driver within human DNA. Lab sciences are working in this territory without knowing it. Psychic Patricia Cori, who channels from the Sirian system, however we might understand that, calls such scientific work "hacking the God code" and believes that it lowers the vibration and leads humans to descend rather than ascend, that is, by serums derived from such hacks.

DNA is the manifestation of the cosmic breath and its knowledge, so meddling with it is dangerous. Some may say that it's the cosmic library, and only certain people have true access to it. Whether Crick and Watson were among the certified, we can only guess. But each human is also equal, and as we move into the realm of 12-stranded DNA, akashic knowledge will be experienced as a source for all. The information flow is mobile, in to out, from its surrounding matrix and electromagnetic fields.

A29. η ενεργοποιησις του δωδεκακλωνικου δεσοξυριβονουκλεϊνικου οξεος = 6284
A30. η εξωτερικη προσωπικοτης του ανθρωπινου οντος = 6284

B29. i energopeoisis tou dodecaclonikou desoxyrivonucleinikou oxeos
B30. i exoteriki prosopikotis tou anthropou

C29. the activation of dodecahedron deoxyribonucleic acid
C30. the outer personality of the human being

Copies of the Divine Genus, Genetic Code
Freedom of Holy Shroud

Simply modified, the phrases "copies of the divine genus" and "genetic code freedom of the holy shroud" emerge. Upon acknowledging 12-stranded DNA, one's divine personality activates. In the future, the human race will experience a transition from identifying as a *Homo sapiens* to a *Divine sapiens*.

In the next phrase, the word *holy* (as in *Holy sapien*) translates to "sacred" in the Hellenic language. A shroud is a covering, to conceal something from being viewed, referring to the husk quality of DNA. The 12-stranded DNA will unveil and free limiting beliefs about former knowledge of the genetic code.

A31. ενεργοποιησις του δωδεκακλωνικου δεσοξυριβοζοπυρηνικου οξεος = 6356

A32. αντιτυπα του θειου γενους και γενετικος κωδικας ελευθερια ιερας σινδονης = 6356

B31. energopeoisis dodecaclonikou desoxyrivozopyrinikou oxeos

B32. antitypa toy theioy genous kai genetikos kodikas elefteria ieras sindonis estin

C31. activation of deoxyribonucleic dodecahedron

C32. copies of the divine genus and genetic code freedom of holy shroud

As a means to begin a further investigation, this approach opens an exciting perspective for understanding the embedded world. The numbers that are listed in the examples above model the potential that gematria and its underlying dodecahedral basis offers; you get to keep the gift of accessing new portals of knowledge and information through this key of accessing cosmic perpetual patterns in nature.

Embrace the process some call *mystery numbers*, guided by subconscious inspiration from angelic and animal frequencies. As we have seen, I decipher their messages in part by gematria, which is also a path of genetic enlightenment. Alphanumeric intelligence is as successful as any other enigmatic technique like Reiki or White Tiger Medicine. This method has played a role throughout history. Samuel Hahnemann diluted homeopathic pills way past Avogadro's number; Dr. William Sutherland accidentally discovered a key to the embryogenic maintenance of the body by therapeutically manipulating cranial bones. James Watson saw DNA first in a Hedy Lamar dance. I learned how to access numbers through letters. After all, letters have a vibration, too. In the next chapter, we will delve deeper into the numerical and gematrical underpinnings of our 12-stranded DNA.

The New Sound Frequencies

Sound Exploration of DNA in the New Age

In my practice, melody, rhythm, and dynamism are active states in producing an emotional response to induce memory activation. In chapter 8, we saw how water translated music from my harp into language using frozen symbology. The music I played to communicate with water included ten notes of my discovery, which I've coined the New Sound Frequencies. These frequencies act as a spiritual attunement to activate our 12-stranded DNA.

Here, we will explore how the New Sound Frequencies were revealed to me: through angelic inspiration and devotion to the dodecahedron form of the DNA molecule. I was divinely led by alpha-numerology, water, and Platonic solids. It cannot be scientifically explained because the 12-stranded DNA begins as a spiritual belief. Its manifestation in physical form is an unknown point in time in the future.

As Rudolph Steiner said, "The human being thinks that he creates intelligence, whereas he only draws intelligence from the universal sea of intelligence." Along the lines of his sentiment, one cannot abstract without the whole existing. In the context of the book, the whole is our 12-stranded DNA, and my abstraction proves that *a* whole exists.

SOLFEGGIO'S HEALING FREQUENCIES PARALLEL THE HELLENIC ALPHANUMERICAL SYSTEM

Solfeggio frequencies, introduced in the eleventh century, were derived from an older six-tone scale discovered by Italian musician Guido d'Arezzo (also see chapter 14). *Solfeggio* refers to *solmization*, the practice of using a syllable to specific notes. For example, the popularized sol-fa method traditionally incorporates the syllable do to the note C, re to the note D, mi to the note E, fa to the note F, and so on. Renowned globally as having healing properties, the nine Solfeggio frequencies, including 174, 285, 396, 417, 528, 639, 741, 852, and 963 Hz, add up to 4,995. Synchronously, the total value of the Hellenic alphanumerical system is 4995. This value is important because it treats the information carried by sound and a word equally in a strict sense. Stringing units of sound creates words, and words are assigned meaning. Therefore, the number 4,995 is shared by the total value of the Hellenic alphanumerical system and the set of Solfeggio frequencies, respectively.

THE DIVINE ENERGY FIELD
Archetypal Frequencies of the Dodecahedron

According to Plato, the dodecahedron symbolizes the creation of the Universe. It was his "big bang," a quiet emanation. Relating the value of the dodecahedron molecule to the 12-stranded DNA, the total sum of the angles in the dodecahedron is 6,480. Using the number 10 as a number that represents universal order (discussed in chapter 14), ten values were selected that 6,480 is divisible by: 108, 144, 216, 324, 405, 540, 648, 720, 810, and 1080. I treated these dodecahedral coordinates as a divine energy field of sound frequencies that indicate one or more archetypal meanings.

The autoharp has 36 strings, giving a variety of harmonic range to explore for sound healing. My harp was custom-designed and strung by professional musician, Mark Torgeson. Torgeson produces a variety of harps that are based on sacred geometries and frequencies, a brand called *Harps for Angels*.[1] My harp features seven out of the ten archetypal frequencies and the tuning pitch is calibrated to note A 432 Hz.

The number 432 isn't a Solfeggio frequency or a New Sound frequency but has a special relationship to photons. The sun's radius is nearly 432,000 miles, rotating on its axis. The speed of light is 186,624 miles per second, which is 432 squared. According to music theory, 432 Hz is mathematically congruent with the Universe and makes the body and its surroundings resonate in a natural way. Italian composer Guiseppe Verdi tuned the note A, known as "Verdi A" to 432 Hz because it was ideal for opera voices.[2] It was adopted as the industry standard due to its universal and spiritual healing components. The number 432 was treated as a center axis for my harmonic Fibonacci frequency sequence, discussed in chapter 8.

The sun doesn't need other stars to shine. It just does and warms us all up while doing it. Similarly, 432 Hz has no association with a specific frequency in hertz. In our universe, the planets revolve around the sun in a metaphorical sense to any frequency pirouetting 432 Hz. Both models activate a cosmic Fibonacci order of vibration, which we perceive as harmony. In this context, harmony can be defined as the integration of time and sound to form DNA. Synchronously, the following translation of the Greek phrase to English "the geometry of harmony is proportionate to time, sound, and DNA" gives the gematrical value of 6485, nearly conjunct by 5 digits from the sum of the angles of the dodecahedron and the 10 frequencies in which the divine energy field was extracted. In summary:

The geometry of harmony is proportionate to time, sound, and DNA = 6485

i geometria tis armonias einai analogi tou chronou tou ichou kai dna = 6485

η γεωμετρια της αρμονιας ειναι αναλογη του χρονου του ηχου και δνα = 6485

The Hellenic word for angel is ἀγγελος, and its gematrical value—remembering that it is also a frequency—is 312. The difference between 432 and 312 is 120. In chapter 7, the 120-cell Form was discussed as the genetic structure of the 12-stranded DNA. Using the Hellenic lexarithmic theory, we see that 432 can also be an angelic frequency and whispers an aspect of the story of our DNA. 432 Hz is not only significant in terms of musical harmony but also introduces a more significant underlying question—what lies beyond our human genetic design when we look past the double helix? Does our DNA contain a marker of a life lurking somewhere between the photons and frequency? These are the questions and themes of this book: that possibly human DNA is the hybrid form of a corporal body and celestial life merged within a unified field, and we are in the exciting beginning stages of remembering our original domain (the basis of recognizing angelic frequency discussed in chapter 15) and composition, our 12-stranded DNA.

Now, let us look at the ten New Sound Frequencies that I've discovered and examine their mystical qualities using gematria. The meanings of frequencies can be decrypted using alpha-numerology, based on their gematria sum. Each frequency is summarized by its geometric properties, including spatial distribution, harmonic behavior, and symbolism in nature.

THE POWER OF NINE

I found that each sound frequency has a special relationship with the number 9: each frequency is divisible by 9, and the sum of each frequency's digits also results in the number 9. Interestingly, all frequencies, except for the odd-number frequency 405, are divisible by 18.

The total sum of the Solfeggio frequencies, the Hellenic alphanumeric system, and the total sum of the angles from each of the Platonic solids can be broken down into a sum of 9:

Solfeggio Frequencies/Hellenic Alphanumeric System: 4995 = 4 + 9 + 9 + 5 = 27 = 2 + 7 = 9
Tetrahedron: 720 = 7 + 2 + 0 = 9
Octahedron: 1440 = 1 + 4 + 4 + 0 = 9

Cube: 2160 = 2 + 1 + 6 + 0 = 9
Icosahedron: 3600 = 3 + 6 + 0 + 0 = 9
Dodecahedron: 6480 = 6 + 4 + 8 + 0 = 18 = 1 + 8 = 9

NEW SOUND FREQUENCIES OF THE 12-STRANDED DNA

Tranquility
108 Hz

The number 36 is an important Pythagorean tetractys, a triangular number that is equal to the sum of the first eight consecutive positive integers:

$$1 + 2 + 3 + 4 + 5 + 6 + 7 + 8 = 36$$

Although the original tetractys was a sacred decad, the Pythagoreans were also interested in the harmonious musical and mathematical relationship of the number eight in nature. Not only does the second order tetractys have a perfect square, but also Pythagorus demonstrated the way eight notes, or the octave, produced a pleasing musical sound according to the vibrating strings with the particular simple ratios of string length. Today, we recognize the relationships the Pythagoreans discovered in music, namely the octave, fifth, and fourth. The ratio of the string lengths also play an essential role in each notes harmonious sound: an octave has a two-to-one ratio, a three-to-two ratio for a fifth, and four-to-three ratio for a fourth. This hidden pattern in sound introduced a broader concept and question. If this common hidden pattern exists in sound and is true universally in all sounds, could it exist universally everywhere? In the context of this book, a water molecule contains a two-to-one ratio of hydrogen atoms to water (also containing eight faces as a polyhedron, or octahedron); Plato authored 36 works; my presented autoharp consists of 36 strings.

The number 108 is equal to 3 times the number 36, while the Pythagorean gematria value of the phrase "the tranquility" is also 108:

$108 = 3 \times 36$

$108 = 1 + 0 + 8 = 9$

$108 = 12 \times 9$

$108 = 6 \times 18$

Η Γαληνη = **108** = > I Galini = > The Tranquility

COMPONENTS

- A dodecahedron contains 5 pentagons. A pentagon contains 5 corners, or vertices. Each vertex point is 108°.

- The chord, or straight-line segment of a circle, of 108° corresponds to the golden ratio, as represented by the equation $2 \sin (108°/2) = \Phi$.

- The number 108 is also a multiple of 432, i.e., $108 \times 4 = 432$; 108 is the first Fibonacci number when the center axis is 432. Although 432 does not naturally appear in the Fibonacci sequence, the concept of using 432 Hz as a "center axis" in the spiral nature of sound frequencies can be explored. The importance of 432 Hz in terms of harmonics and its potential in doubling and halving processes is particularly noteworthy. When 432 is halved, there is a progression through 216, 108, 54, and then 27 where the sequence of whole numbers stops. The number 27 is significant in geometric and spatial contexts—it is the number of spheres needed to form a cube ($3 \times 3 \times 3$). This geometric representation could provide insight into the spatial properties of sound. Considering a Fibonacci analog that begins with a different starting point, highlighting 432 Hz as the center axis, the pattern would begin as 108, 108, 216, 324, and 540. The growth pattern reflects a scale phenomenon found in nature, which may suggest a deeper connection between Fibonacci numbers and sound frequencies. Starting with the recursive sequence based on 108 and continually adding it to itself (as with any two numbers), the ratio of successive terms will stabilize and converge on the golden ratio, which is approximately 1.618.

The Divine
144 Hz

The number 144 is the perfect square of the mystical number 12, the twelfth number in the Fibonacci sequence. It was the vertex angle of the triangular pediments of the ancient Hellenic temples. The number 144 is woven into intricate patterns and represents universal life such as the geometric design known as the Flower of Life and the Shri Yantra. The number 144 is equal to the gematria sum of the phrase "the divine:"

$144 = 4 \times 36$

$144 = 1 + 4 + 4 = 9$

$144 = 16 \times 9$

$144 = 8 \times 18$

Θειον $= \mathbf{144} = >$ Theion $= >$ The Divine

COMPONENTS

- A dodecagon, evident in a cross section of DNA, has 10 internal angles that are 144° each. There are 10 golden triangles, an isosceles triangle, depicting the phi ratio.
- The sum of the areas formed on the legs with the numbers 144 and 108 equals the hypotenuse, 180 ($144^2 + 108^2 = 180^2$). The angle sum in a triangle is 180 degrees.
- Since 144 is one-third of 432, we can treat 144 Hz as the fundamental frequency ($n = 1$) and 432 Hz as the second overtone and 3rd harmonic* ($n = 3$).

Polytopes are geometrical figures bounded by hyperplanes that connect multiple dimensions. In a dodecahedron modeled as a 120-cell, hyperplanes intersect at angle of 144°.[3]

---

*The fundamental tone of a sound frequency is a unit of its underlying basic waveform. Sounds are distributed by a number of different waveforms and frequencies in nature. When two separate waveforms meet, the superimposed resulting wave is called a standing wave. Overtone refers to higher frequency standing waves that occur simultaneously, stemming from the fundamental tone. Harmonic is a term that encompasses all the overtones, representing them as whole-number multiples of the fundamental tone.

Breath
216 Hz

The number 216 is mentioned several times in Plato's texts and especially in the *Laws*, where Plato mentions that the minimum gestation period of an embryo should be 216 days. The number 216 is a perfect cube, and, in fact, it is the cube of the number 6. By adding a zero to the numerical order of 216, we obtain 2,160 years, which represents an astrological age. The gematria value of the phrase "the breath" is also 216:

$$216 = 6 \times 6 \times 6 = 6^3 = 2 \times 108$$
$$216 = 2 + 1 + 6 = 9$$
$$216 = 24 \times 9$$
$$216 = 12 \times 18$$
$$\text{H } \pi\nu o \eta = \mathbf{216} = > \text{I Pnoi} = > \text{The Breath}$$

COMPONENTS

- The number 216 is the second Fibonacci number when 432 is the center axis. The number 216 is grouped as a family with 288 and 360. In a right triangle with sides of 216 and 288, the sum of the areas formed on the legs of the triangle equals the length of its hypotenuse, 360 ($216^2 + 288^2 = 360^2$).
- The frequency 288 Hz is the first overtone following 144 Hz. The number 360, a full circle or pi, means completion.
- A dodecahedron contains 20 vertices, and each vertex is 108°. The sum of all vertices is 2160° (20 × 108°), another order higher than 216 Hz.

Divine Harmony, Birth, Pure
324 Hz

The number 324 is the perfect square of the number 18, while the number 18 is equal to the number produced by the sum of the numbers 3, 6, and 9. Serbian physicist Nikola Tesla was fascinated with the numbers 3, 6, and 9. Although his theories were grounded in empirical observation, his wonder with these three digits reflected his passion to discover the hidden and rhythmic patterns of the unseen world. In the same context, the gematria presented in this book is a means to empirically observe the geometry of words to uncover the mysterious connection between ideas and concepts. Indeed, the following four relations prove it to us:

$$3 + 6 + 9 = 18$$
$$324 = 18 \times 18 = 18^2$$
$$324 = 3 + 2 + 4 = 9$$
$$324 = 36 \times 9$$

Furthermore, we notice that the numbers of the phrases, "divine harmony," "birth" and "pure," are equal to the number 324. The interpretation of this equality is that humans are born with divine harmony and at birth are pure, just like nature and creation.

Θεικη αρμονία = γεννηση = αγνος = **324** = > Theiki armonia = genesis = agnos = > Divine Harmony = Birth = Pure

COMPONENTS

- The number 324 is the third Fibonacci number of 432.
- The dodecahedron contains 3 angles at each vertex. This is where 3 pentagons meet. The sum of the angles at each vertex is 324° (3 × 108°). It is also the vertex point where 3 pentagrams meet.
- The sum degree of all edges of a dodecahedron is 3,240° (30 edges × 108°), a numerical order higher than 324 Hz.

Divine Evolution
405 Hz

The sum of the first nine consecutive positive integers is the number 45:

$$1 + 2 + 3 + 4 + 5 + 6 + 7 + 8 + 9 = 45$$

Multiplying the number 45 by 9, we get the number 405, which is the number of the ancient Greek prefix ευ (eu) meaning "good" (eulogy, euphoria, etc.), while the number 405 is also the number derived from the sum of the phrase divine evolution (Theia exelixis). The following equalities demonstrate the connection to the number 9:

$405 = 4 + 0 + 5 = 9$

$405 = 45 \times 9$

Eu = Θεια εξελιξις = **405** = > Eu = Theia Exelixis = > Good = Divine Evolution

COMPONENTS

- The number 405 represents a pentagonal pyramid. A pentagonal pyramid has a pentagon base and five triangular faces that meet at the apex. When you look at the flat net of a pentagonal pyramid, it is the shape of a five-pointed star. A flat net is a pattern laid flat that can be folded to form a three-dimensional figure. It can be seen as the top portion of the icosahedron.

A pentagram, one of the most recognized symbols in occultism, is a five-pointed star containing 10 isosceles triangles, 5 golden triangles and 1 pentagon. This arrangement forms 5 obtuse angles and 5 acute angles within the star, symbolizing the concept of sacred duality. Among its rich history to world culture, the pentagram was a symbol of interest to Pythagoras. Comprised of the "five geometrically perfect A's," the pentagram contains a regular pentagon and five golden triangles, which reveals not just one golden ratio, but multiple occurrences of phi. The pentagram contains nested pentagons and numerous possibilities for the golden ratio, which symbolized that every number-shape had a hidden meaning for the Pythagoreans.

Along with the aspects of pentagon symbolism being associated to perfection and sacred geometry, the pentagram reminds me of angel symmetry. By perspective, even though we do not necessarily see a Fibonacci spiral or golden ratio in nature or culture, the patterns of the golden ratio and sequences still exist among us and beyond what meets the eye. When we do the math behind some of the elusive geometric shapes, the confirmation of the golden ratio is possible. This serves as a metaphorical example, suggesting that angels exist among us, even if they are not always visible. After all, angels are accessible to human beings through sacred words, nature's geometry, and sound frequencies. Ten isosceles triangles compose the pentagram and ten is associated with the angelic realm, which was discussed in chapter

14. From the saying "as above, so below," the number 10 depicts an order of magnitude of the cosmos and the angelic realm.

The Divine Spiral
540 Hz

The number 540 is equivalent in gematria to the phrase, "the divine spiral." Σπειρα means "spiral." Another translation for the Greek word for *spiral* is "galactic." The concept of the divine spiral empowers our dreams, imagination, inspirations, thinking, ideas, and psychic visions.

$540 = 5 \times 108$

$540 = 15 \times 36$

$540 = 30 \times 18$

$540 = 5 + 4 + 0 = 9$

$540 = 60 \times 9$

Σπειρα θειον $= \mathbf{540} => $ Spira Theon $=>$ The Divine Spiral

COMPONENTS

- The number 540 is the fourth Fibonacci number starting at 432.
- The sum of all vertices in a dodecahedron is 540° ($5 \times 108° = 540°$).
- The number 540 is the twelfth number in the decagonal pattern, which means 540 is able to arrange itself into a decagon.

Cosmic Harmony
648 Hz

The number 648 is equal to three times the number 216 and equal to the gematria sum of the phrase, "cosmic harmony:"

$648 = 3 \times 216$

$648 = 18 \times 36$

$648 = 6 + 4 + 8 = 18 = 1 + 8 = 9$

$648 = 72 \times 9$

Η κοσμικη αρμονια $= \mathbf{648} => $ I Kosmiki Armonia $=>$ The Cosmic Harmony

COMPONENTS

- The total sum of internal angles of a dodecahedron is 6,480°, which corresponds to its symmetrical structure.
- The dodecahedron also has a sum of vertices of 6,480° (12 faces × 540°, and 20 vertices × 324°).
- The difference in frequency between 648 Hz and 1080 Hz is 432 Hz, which may suggest a relationship between these frequencies.

Mentality
720 Hz

The number 720 has the same gematria for the word *mind*, or mentality. This concept evokes the notion of the universal mind that shaped the cosmos, mirroring how our mind shapes our reality.

$$720 = 5 \times 144$$
$$720 = 20 \times 36$$
$$720 = 40 \times 18$$
$$720 = 7 + 2 + 0 = 9$$
$$720 = 80 \times 9$$
$$Νους = 720 = > Nous = > Mind$$

COMPONENTS

- The 120-cell (dodecahedron) contains 720 pentagonal faces, and the 600-cell (icosahedron) contains 720 edges.
- The sum of all vertices in a tetrahedron is 720° (4 faces × 180° = 720°).
- The internal angles of a golden triangle are 72°–72°–36°.
- The frequency 720 Hz is the fifth harmonic within 432 Hz. There are five total units, describing each side of a pentagon.

Friend, Brother, Paraclete
810 Hz

The number 810 is equal to the gematria of the Greek words for "friend," "brother," and "Paraclete," while it also equals twice the number 405, as well as the product of the numbers 18 times 45. Indeed:

$$810 = 2 \times 405$$

$810 = 18 \times 45$

$810 = 8 + 1 + 0 = 9$

$810 = 90 \times 9$

Φιλος = αδελφος = Παρακλητος = **810** = > Filo = Adelfos = Paraklitos = > Friend = Brother = Paraclete

COMPONENTS

- There is a unique aspect of the number 810. It is derived from the sum of all numbers divisible by 360:

$$180 + 2 + 120 + 3 + 4 + 90 + 5 + 72 + 6 + 60 + 8 + 45 + 9 +$$
$$40 + 10 + 36 + 12 + 30 + 15 + 24 + 18 + 20 + 1 = 810$$

What is significant about the number 360? The number 360 can denote multiple meanings. My interpretation of the number 360 is associated with spin or aliveness, completion (portrayed by the fullness of a sphere), a cyclical trait. The relationship of 36 and 0 can also resemble a "divine tetraktys."

Holy Spirit
1080 Hz

Since each of the eight equal interior angles of a regular octagon is equal to 135°, the sum of the interior angles of a regular octagon is equal to 1080°:

$$8 \times 135 = 1080$$

The number 1080 is equal to five times the number 216, as well as three times the number 360, and is equal to the sum of the numbers of the three words that make up the phrase Το Αγιον πνευμα (To Agion pnevma) which translates to the Holy Spirit:

$1080 = 5 \times 216$

$1080 = 3 \times 360$

$1080 = 30 \times 36$

$1080 = 60 \times 18$

$1080 = 1 + 0 + 8 + 0 = 9$

$1080 = 120 \times 9$

Το Ἅγιον πνευμα $= \mathbf{1080} = >$ To Agion pnevma $= >$ The Holy Spirit $= 1080$

COMPONENTS

- In a dodecahedron, the total number of vertices is $10,800°$ (20×540), which is one order higher than 1080 Hz.
- The 6 triangles in the sum of the angles in David's Star equals $1080°$.

12-STRAND FREQUENCIES ENTER THE FOURTH DIMENSION AND INFORM NEW THERAPIES

The 12-strand frequencies—108 Hz, 540 Hz, 648 Hz, and 1080 Hz—seem to resonate with the four-dimensional cell structure of DNA: the 120-cell containing 120 dodecahedra, 600 vertices, 720 pentagons, and 1,200 edges, can be reduced by an order to 12, 60, 72, and 120 and factored by 9. These sound and geometric synchronicities indicate an energetic bridge between our genetic imprint and celestial ancestry. They represent the harmony of cosmology, nature, and humanity.

My theory is an inquiry into deep DNA, beyond Watson-Crick and attuned to harmonizing with nature. The new sound frequencies' whole-body vibrational effect on human health is a path of future exploration. Regarding future prospective explorations of this book, these frequencies could be key to locating therapeutic DNA sequences, similar to Dr. Temple's method of developing his own DNA music. By using parallel theories, we can understand emotions within our psychological mind and knowledge within our intellectual mind, using a new spectrum of sound.

This book accumulated available knowledge about DNA, sound, Platonic solids, geometry, water, and psychology and integrates universal law to propound the difference between a double helix and a dodecahedral DNA molecule. By applying gematria, we can explore and potentially resolve cosmological and theoretical mysteries, providing insights and circumstantial evidence for further investigation. Despite the propelling number of genetic research, advancements, and achievements in the twenty-first century, modern science puts a limitation on the cosmological and physical entities by treating them in the same paradigm. How can the laws of cosmology follow the same laws of matter (or physical science) when the laws of matter are separate from the laws of biological sciences, psychology, or intellect? Each level exists with its own parallel laws and distinguishing its corresponding parameters is vital to deepen our understanding in the different levels of life in the Universe.

As we evolve our perspective and orientation towards science, technology, and wellbeing in this era, perhaps one way to respect the parallel laws of cosmology and matter as it relates to human beings is by incorporating parallel logic and theories. In chapters 16 and 17, I show how gematria facilitates arithmetics to compare compounded words and phrases with the meanings of the messages. This method demonstrates cosmological intelligence that informs a human's physical, biological, psychological and intellectual brain. While the principle of observing the relationship between numbers and nature emanates from one of the first civilizations, the applications of gematria could be a considerable mathematical approach and potential application used to reach new realities of the cosmological and physical world, respectively.

∽০∾

While birds rely on their strong wings to fly, humans can harness the power of their will to soar, experiencing the world with newfound determination and hope. Upon completing the Ironman in Australia and placing in the top 10 percent of my age group, I returned home with a renewed mindset and a heightened sense of belief. If I can build on my grief, make a jump like I did with the race, and transform into the Dragon, what is stopping me from doing the same with my career?

How do I share this feeling of happiness, livelihood, and enrichment with others? I remembered how I felt when I started to feed my body natural, whole ingredients; how much my focus and energy increased. I looked down at my cup that I was drinking from. It was probably the three hundredth smoothie that I had made using almond milk, spinach, golden raisins, almond butter, banana, and chia seeds. It changed my life at the time. If it did for me, then it could for another. And even if it was just one person that it touched, then it would have accomplished its mission.

To be an entrepreneur, one must have a sense of adventure and the willingness to take risks. Moving forward, I decided that I must start with my green smoothie that I built for my triathlon training, a legacy in honor of my father.

Without my father around for approval and my current, my family's concerns and the ongoing challenges of my divorce settlement, I came to realize that no one was going to teach me how to fly. I had to simply push myself out of the nest and expect to work it out on my own. My pursuit for evolution is both my own right and responsibility.

It was a terrifying and confronting move. However, it would be even more scary to imagine what life would look like if I stayed. How else would I have been able to fall in love again, or learn greater dimensions, or write this book?

I gave my team, department, and company a month's notice, tying up loose ends, training new hires, and preserving the professional relationships that I cultivated over the years. My department wrote the loveliest group card and threw me a party during the last week. It felt as if this chapter of my life was experiencing a death and I was alive to see how it was remembered. I received gifts, tears, hugs, plants, and salutations. This experience served as a crucial turning point, enabling me to solidify my moral and spiritual stance on health, science, and safety.

Epilogue

I have discussed pattern recognitions interlinking DNA and angelic frequencies. Using gematria, I highlighted its genetic geometry and identified an exclusive set of 12-strand sound frequencies. Then I interpreted its vibration using water consciousness. We explored activating unconscious memories of our ancestors and two-way communication through lucid dreaming. Healing ancestral relationships means going back to the land to compost grief and extending our synchronization to encompass animal and angelic frequencies.

What's next? Without action and having a practice, we cannot evolve beyond the realms of conventional physics, chemistry, and clever mathematical models. To alter our DNA we have to change the way we think, even as other animals in nature do. Similarly to the slime mold mentioned in chapter 3 and scaled to a human being's version of time, we decide as a collective when it is time to explore our environment and when to move to the left or to the right. The mechanism in which information is transferred inside the slime mold and the way it contracts into the direction that it is drawn to go is mirrored in the way humanity has physically moved in time and space seeking new ways to live. Even by physical manipulation, scientists are fascinated by the choices that the slime mold is influenced to make when it is poked by a section of the membrane. It leaves them at a philosophical junction: maybe intelligence is not that difficult, or, perhaps we should redefine what we mean by intelligence.

We ask ourselves using the intelligence encoded in the slime mold's primitive model as an example, what mysteries do we seek within ourselves? How can we unlock new understanding in our "brainless intelligence" to re-emerge with an innovative gateway of potential, understanding, and connection of our physical life and cosmological background? Activating our 12-stranded DNA may hold the key. By parallel, this book means to initiate a chemical signal to our mind and body by beginning with a physical indication. The process in which a change in the slime mold's direction is based on the physical conditions of their environment before the occurrence of a chemical signal is a paradigm of our collective human experience.

However, the moral of this book is not just about changing and altering strands of our DNA as a psychic meditation. It is about the way we feel about ourselves, the way human beings treat each other and heal through generations of grief. Grief blankets memory; it is the reason why we've forgotten how we've come to this point, why we live and continue to walk this planet, recycling sorrow through zombie mindsets drowned in density. However, the time has come to believe in the goodness and healing energy of our souls. Our souls are evolving, so we can go deeper.

The history of humanity has been wrapped in darkness for many generations. Yet, we are experiencing a mysterious and precious silence, a silent sound bath before dawn. Remember the sensation when you feel a subtle shift, in mood or weather or fortune. Something novel is aspiring to be known. Recall the sweet moment before a smile or receiving a warm welcome, or a loved one opening their arms for a hug. It's the middle ground, the space in between, and how we decide to perceive it.

Our ancestors, elders, and ascendants—yes, ascendant, our descendants too—chanted, prayed, and bowed to the land in honoring a phenomenon that cannot be fully understood. Yet they accepted and turned to it. They honored existence as a miracle. In the present world, what makes a miracle? In science, it can be a reversal of death, one reason that legendary resurrections and rainbow bodies remain a big deal. But miracles don't evade or elude death; they aren't of the quality of death, they are sourced from life, being, spin. In spiritual consciousness, a miracle is the awakening and individuation of human sovereignty. Miracles are sparks of genesis, dawns after darknesses. A new expression of DNA is awakening moment by moment in every cell we inherit and bequeath. The activation of DNA memory is the activation of soul memory, where atman meets genome. At the aha moment, we feel our inner knowing come alive and mysteries turn into facts. Miracles aren't an exaggeration or violation of life, they *are* life.

Miracles are central to our soul's evolution, too. They meld the present, past, and future to hope. The human species fast-tracked the manifestation of the material world through technology: cyborgs, humanlike robots, DNA editing formulas, cloned living organisms—hiding a natural miracle behind mechanical distractions. Of course, even synthetic biology is biological, offshoots of humans as much as hives are offshoots of bees and dams are offshoots of beavers. Can we also purify our environment and waters and support natural animal habitats and cycles? The present waning of nature reflects the way we think about ourselves.

We can each begin by addressing our grief. Even if you are not actively grieving, you are carrying grief: personal grief, tribal grief, ancestral grief, collective grief, Gaia's grief. Little shifts in our thoughts then shift the way we perceive our environment, which shifts the way it "perceives" us. If by our existence we are a miracle, then by conducting ourselves with clarity and sincerity, we can extend the miracle and magic. When we acknowledge such a principle—sympathetic, telekinetic, and transpersonal—we reclaim our innate ability to perform miracles. They may seem miniscule at first, but a miracle is a miracle and, if you can perform one, you can perform others. We might begin by cleaning our biomes to begin renegotiating our energy contract with our own planet. My book is not an eco-primer. It is a way to awaken your consciousness so that eco-stewardship and innovation come naturally.

It's challenging to see something clearly when our mind is clouded with emotions, trauma, and energy that is not ours. Sound healing is one foundational stepping stone. Why sound? Sound attunes the brain, mind, light, the auditory canals, the steady-state hum of our existence, and the rest of our senses by vibration and resonance. Sound and light inscribe new content in our genes—they can't help but do so, once entrained and canalized, because genes *are* sound and light. A gene is a miniature representation of the universe, retaining hundreds of years of memory and connecting us to our ancestors. When you look at telescopic images of shimmering galaxies, you are looking at the filaments and spirals of genes writ large, and vice versa.

Through genes and aetheric fields, we literally remember our ancestors and honor their lives by applying their wisdom and experience as distilled knowledge to our future. We don't even have to see them if we accept our mutual healing, and that's okay. A time will come when each person's unique evolution will reach a point of conscious recognition.

Before we can honor our ancestral contracts, we need to clear the emotional space of generational grief. To shift our relationship with that grief, we begin by accepting—truly accepting—that we don't want to lumber on in guilt, fear, or rage anymore, meaning that we recognize the ways in which we *did* want to, leading to our main spiritual enemy: self-sabotage. When we heal ourselves, we can impart a universal sense of belonging and change patterns backward and forward through many generations.

You don't have to believe that our ancestors and angels are present for them to be so. I promise, they are. We cannot change what happened in the past, but we can cultivate a sacred mind-frame in the now and future to recover from generational trauma. "Sacred" means taking an ordinary space and making it your own interpretive heaven, a vibration or portal (depending on which clairsentient sense you use, connecting your mind to your soul). Creating a sacred communion empowers you to see that you and everything else are inseparable. From practicing this mental adjustment, you are magically rewarded in a moment, a day, a week, a year, or a lifetime. The adjustment will cut through time and show you why "wait time" doesn't matter. I'm not saying that my methods work for you. Alchemize your own creative approach to shifting your vibration and composting grief. When you shift, the course of your life changes, and so do the courses of those around you. We leave a lovely nest for our next life's incarnation and future generations.

As we explore a new politics of consciousness, we shed old partisan skins. We may never fully disentangle ourselves from the complex choreography of material events, but we can evolve with a quality of being that is beyond what is scientifically rational. Healing consciousness breathes in the mystery of life and its surprises, and exhales away from our yearning to "solve" these riddles. It is a high time in our thinking about the world and trusting our own heart, dissolving any vestigial shame, to douse anger with this psychic ambrosia so that we can access our vitality. While most of the cosmic substance is dark, our temporary sentient bodies shine with significance.

Our DNA *wants* to be addressed. It *wants* to be taken care of. It *wants* to be activated. More people are starting to realize it. I recently talked with a friend who was expressing her frustration about something and she said to me, "I don't want to say what I really want to say out loud because I don't want my cells to hear it."

My practice shifts the sound that originates in our thoughts, the vibrant quality in our heart, and the internal rhythm that we project to the world so that we can begin to evolve with greater awareness of the sounds, information, and intentions from our external environments. Think about it. Any given organism (whether it is organic or artificial) functions based on the way it is wired within. The beauty of being human is that we have the choice of rewiring our system—thanks to frequency—so that we can continue developing new limits in understanding life. Human beings share a cosmic intelligence; we continue evolving over time. What path will you choose to take? What form do you see yourself evolving into? What does your heart wish to co-create with your wondrous and miraculous vessel?

In *The Alpha Numeric Intelligence of the Hellenic Language*, readers may be astounded to witness the profound interconnectedness of the following distinct concepts:

Watson + Crick = Gouotson + Krik = Γουοτσον + Κρικ = **1313**

The helix of the DNA = I elix tou nti en ei = Η ελιξ του ντι εν ει = **1313**

The dodecahedral = O dodekaedrikos = Ο δωδεκαεδρικος = **1313**

Gnomonic = Gnomonikos = γνωμονικος = **1313**

Genetic Engineering = Tzenetik entziniaringk = τζενετικ εντζινιαρινγκ = **1313**

Reflecting on these shared gematria values, what inspirations arise within you?

Notes

INTRODUCTION

1. Watson, Gann, and Witkowski, *The Annotated and Illustrated Double Helix*, 25–26.
2. Braun, Tierney, and Schmitzer, "How Rosalind Franklin Discovered the Helical Structure of DNA," 140.
3. Hubbard, *Science, Power, Gender: How DNA Became the Book of Life*, 269.
4. Watson and Crick, "Molecular Structure of Nucleic Acids," 737–38.

CHAPTER 1. WHAT IS 12-STRANDED DNA?

1. Antonio et al., "Single-Molecule Visualization of DNA," 832–37.
2. Liu, Liu, and Zhang, "Compositional Bias," 130–40.
3. "Science Historian," University of Manchester Website.
4. Genesis 1:27, NIV.

CHAPTER 2. THE ART OF MEMORY AND GENERATIONAL HEALING

1. Koch, "Activating False Fear Highlights a Memory's Neural Trace," November 20, 2022.
2. Penfield, *Mystery of the Mind*.
3. Adler, "The Sensing of Chemicals by Bacteria," 40–47.
4. Darwin, *The Descent of Man, and Selection in Relation to Sex*.
5. Tyng et al., "The Influences of Emotion on Learning and Memory," 1454.
6. Gao et al., "The Neurophysiological Correlates of Religious Chanting," 4262.

7. Harmony, "The Functional Significance of Delta Oscillations," 83.
8. Jiménez et al., "Principles, Mechanisms and Functions of Entrainment," 20210088.

CHAPTER 3. DEVOLVING TO EVOLVE

1. Johnson, *Emergence*, 11–23.
2. Ha, Morrow, and Algiers, "Slime Molds."
3. Smith, *Wealth of Nations*, 35.
4. Berkeley, University of California, "What Has Science Done for You Lately?"
5. Kopi Luwak Direct, 2021.
6. Gommans et al., "RNA Editing."
7. Carter, "Octopus DNA Is Not of This World, Numerous Researchers Conclude," *Anomalien* website.
8. Kaliman, "Epigenetics and Meditation," 76–80; Chaix et al., "Differential DNA Methylation," 36–44; Mendioroz et al., "Telomere Length Correlates," 4564; Abomoelak et al., "Cognitive Skills and DNA Methylation," 1214; Verdone et al., "Epigenetic Effects of Meditation," 339–76; Dasanayaka et al., "Association of Meditation with Telomere Dynamics," 1222863; Venditti et al., "Molecules of Silence," 1767; Kripalani et al., "The Potential Positive Epigenetic Effects," 827–32; Dema et al., "Mindfulness-Based Therapy on Clinical Symptoms and DNA Methylation," 2717–37; Radon, "Meditation Practices on Epigenome Dynamics," 17–24.

CHAPTER 4. THE GENETIC ARCHETYPE AND THE DODECAHEDRON FORM

1. Browne, *The Garden of Cyrus*.
2. Plato, *Phaedo*.
3. Plato, *Republic*.
4. Marrin, *Universal Water*.
5. Bandyopadhyay, *Nanobrain*, 19.
6. Spellman, "Symbolic Significance of the Number Twelve," 79–88.
7. Henig, *Religion in Roman Britain*, 128.
8. Abbatantuouno, "Armand Spitz."
9. "Spitz Junior Planetarium," National Museum of American History.
10. Abbatantuono, "Armanda Spitz."

11. Doughtery, "A Brief History of UCSF Chimera."

12. Marrin, *Water and Nature's Geometry.*

13. Hoelz, Nairn, and Kuriyan, "Crystal Structure of a Tetradecameric," 1241–51.

14. Deshmukh, "Consciousness, Awareness, and Presence," 144–49.

15. Plato, *Timaeus.*

16. A line from William Blake's *The Lamb.*

CHAPTER 5. REVEALING OUR 12-STRANDED DNA BY PYTHAGOREAN GEMATRIA

1. Smith, "Bioelectrodynamics and Biocommunication," 102.

2. "Science Historian," University of Manchester.

3. Adrados, *A History of the Greek Language*, Prologue.

4. Alphabet (Early Greek), Brown University.

5. Andriotis, Nikolaos P., *History of the Greek Language.*

CHAPTER 6. MOLECULAR GEMATRIA

1. Tosenberger, et al., "A Conceptual Model of Morphogenesis," 283–94.

2. Bateson, "The Corpse of a Wearisome Debate," 2212.

3. Mandal, News Medical Life Sciences, "What Is Junk DNA?"

4. Feschotte and Pritham, "DNA Transposons," 331–68.

5. Zhang et al., "Rapid Evolution of Protein Diversity," 679–90.

6. Torgeson and Hernandez, "Evolutionary Signatures in Non-coding DNA," 115–124.

7. Brown, Baker, and RSF Research Scientists, "The Morphogenic Field is Real."

8. Montagnier et al., "Transduction of DNA," 106–12.

9. Montagnier et al., "DNA Waves and Water."

10. Gariaev, et al., "Materialization of DNA Fragment in Water," 1–56.

11. Gariaev, et al., "DNA as Basis for Quantum Biocomputer," 25–46.

12. Schempp, "Quantum Holography," 12.

13. Gariaev and Pitkänen, "Model for the Findings about Hologram."

14. Feynman, "Plenty of Room at the Bottom."

15. "DNA Storage: Revolutionary Technology," Sciences Sorbonne Université.

16. Goldman et al., "Toward Practical High-Capacity," 77–80.

17. Taylor, "Volume of Data/Information Created," Statista
18. Grossinger, *The Return of the Tower of Babel*, Substack.

CHAPTER 7. WATER MEMORY AND MYSTERIES

1. Dayenas et al., "Human Basophil," 816–18.
2. Zheng and Pollack, "Long-Range Forces."
3. Klimov and Pollack, "Visualization," 11890–95.
4. Franklin and Gosling, "The Structure of Thymonucleate Fibres," 673–77.
5. Souam, "The Schäfli Formula," 31–45.
6. Besson, et al., "The Adenovirus Dodecahedron: Beyond the Platonic Story," 718.
7. Zhang et al., "Spectroscopic Observation," 1467–73.
8. Zhang et al., "Spectroscopic Observation," 1467–73.

CHAPTER 8. WATER FREQUENCIES

1. Whiteley et al., "Correlating Dynamic Strain," 3386.
2. Quaglia, "Why Scientists Are Turning Molecules into Music."
3. Subagia, Sena, and Suta, "Comparative Study of Water."
4. "Dr. Masaru Emoto and Water Consciousness," The Wellness Enterprise website.
5. Wu et al., "Bach is the Father of Harmony."

CHAPTER 9. SOUND ALCHEMY

1. "Mysterious Ancient Temples," Interesting Engineering.
2. Lynch, "Tumors Partially Destroyed with Sound Don't Come Back," Michigan Engineering News.
3. Khan, *Mysticism of Sound and Music*, 10.
4. Kurek, *The Sound of Beauty*, 153.
5. Didgeridoo, Duke University Musical Instrument Collections.
6. Meymandi, "Music, Medicine, Healing," 43–45
7. Meymandi, "Music, Medicine, Healing," 43–45
8. Burns, "Tibetan Singing Bowls."
9. Meessen, "Virus Destruction Resonance," 2011–52.
10. "Why Do Cats Purr?" Scientific American website.
11. ABC Science website, "The Science of Meditation."
12. Goetz, "Jean-Martin Charcot," 475–78.

13. Charcot, "Vibratory Therapeutics," 821–27.

14. Charcot, "Vibratory Therapeutics," 265.

15. Campbell, "Vibroacoustic Treatment and Self-Care."

16. Campbell, "Vibroacoustic Treatment and Self-Care."

17. Zhou et al., "Osteogenic Differentiation," 12–25.

18. Prè et al., "High-Frequency Vibration."

19. Hoffman and Gill, "Applied Vibration," 274–84.

20. Underwood, "Sound of Mom's Voice Boosts Brain Growth in Premature Babies."

21. Kim et al., "Overexpressed Calponin3," 48–62.

22. Gilman, "Joint Position," 473–77.

23. Kantor et al., "Potential of Vibroacoustic Therapy," 3940.

24. Kantele et al., "Effects of Long Term," 31–37.

25. Lundeberg et al., "Pain Alleviation by Vibratory Stimulation," 25–44.

26. Ko et al., "Effects of Three Weeks," 99–105.

27. Cerciello et al., "Clinical Applications," 147–56.

28. Cochrane, "Effectiveness of Using Wearable Vibration Therapy," 501–9.

29. Sharma et al., "Investigating the Efficacy of Vibration Anesthesia," 966–71.

30. Sandri et al., "Foxo Transcription Factors," 399–412.

31. Ren et al., "Low-Magnitude High-Frequency Vibration," 979–89.

CHAPTER 10. DRAW BREATH AND DNA CONSCIOUSNESS

1. "Croatian Freediver," The Indian Express website, November 20, 2022.

2. Hunt, *Your Spark Is Light: The Quantum Mechanics of Human Creation.*

3. Paul, "Radiant Zinc Fireworks Reveal Quality of Human Egg;" Koumoundouros, "The Biological Fireworks Sparked by Fertilization."

4. Peacock, "Synesthetic Perception," 483–506.

5. Galeyev and Vanechkina, "Was Scriabin a Synesthete?" 357–61.

6. Abhang et al., "Chapter 2-Technological Basics of EEG," 19–50.

7. Dossey, *Healing Words.*

8. "The Science Behind the Wim Hof Method," Wim Hof Method website, November 20, 2022.

9. "The Science Behind the Wim Hof Method," Wim Hof Method website, November 20, 2022.

10. Martin et al., "Functionally Distinct Smiles," 3558.

CHAPTER 11. DREAMTIME AND THE ANCESTRAL REALM

1. Grossinger, *Bottoming Out the Universe*, 85.
2. Steiner, *Sleep and Dreams*, 128, 117.
3. Wolkstein and Kramer, *Inanna: Queen of Heaven and Earth,* 66.
4. Konkoly et al., "Real-Time Dialogue."
5. "Friedrich August Kekulé, Famous Scientists.
6. Feder, Newsela. Dmitri Mendeleev.
7. Feuer, The Dreams of Descartes.
8. Tatera, "Breakthroughs and Innovations," The Science Explorer website.

CHAPTER 12. COMPOSTING GRIEF

1. McMorrow, *Grounded*, 88.
2. "All the Compost Creatures: Levels 1, 2 and 3." Planet Natural Research Center.
3. "All the Compost Creatures: Levels 1, 2 and 3." Planet Natural Research Center.
4. Ningthoujam and Singh, "Possible Roles of Cyclic Meditation," 768031.
5. Alonso et al., "Maladaptive Intestinal," 163–72; Alvery et al., "Gut-Derived," 480–89; Cogan et al., "Norepinephrine," 1060–65; Freestone et al., "Involvement of Enterobactin," 39–43.
6. Maeder et al., "Temporal and Spatial Dynamics," e0263618.

CHAPTER 13. ANIMAL MEDICINE AND HOLISM IN THE NEW AGE

1. Grossinger, *The Return of the Tower of Babel*, 281.
2. Gurdjieff, *Views from the Real World*, 244.
3. Early, "Power of the Animals."
4. Sharp, Chi Flow, "12 Animals of Xingyiquan."
5. Caldwell, "Our Forests are Made of Salmon," North Coast Land Conservancy website.
6. "12 Facts About Otters," U.S. Department of the Interior website.
7. Lewis, *Comic Effects*.

CHAPTER 14. DIVINE CODES, HUMAN ORIGINS, AND ANGELIC FREQUENCIES

1. González-Wippler, *The Kabbalah and Magic of Angels*, 110.
2. Kaler, "Arcturus."

3. Navarro, "The Extragalactic Origin of the Arcturus Group," L43–46.

4. Allen, *Star Names and Their Meanings*, 101.

5. Cayce, "Arcturus: A Compilation of Extracts from the Edgar Cayce Readings."

6. Gardener, The Secret Art of Alchemy.

CHAPTER 15. ANGELIC FEROCITY

1. Agrawal, "Electromagnetic Wave Absorption," 48–65.

2. Friedlander, *Recentering Seth*, 185.

3. Grossinger, *The Night Sky: Soul and Cosmos*, 312.

4. Sheldrake, Smart, and Avraamides, "Automated Tests."

5. Brophy, *Don't Never Forget*, 285–89.

PART 3: INTRODUCTION

1. Cowley and Lawler, Paul Valéry, vii.

CHAPTER 16. GEMATRIA MEDITATIONS

1. "Most Proofs of Pythagoras' Theorem," Guinness World Records.

CHAPTER 17. THE NEW SOUND FREQUENCIES

1. Torgeson, "Harps for Angels."

2. Morgan, Amanda, "Music Theory and the History of 432 Hz."

3. Coxeter, *Regular Polytopes*, 292–93.

Bibliography

Abbatantuono, Brent. "Armand Spitz—Seller of Stars." University of Florida. International Planetarium Society, 1995.

ABC Science. YouTube website. "The Science of Meditation." Abhang, Priyanka A., W. Gawali Bharti, and C. Mehrotra Suresh. *Introduction to EEG-and Speech-Based Emotion Recognition.* 19–50. Academic Press, (2016).

Abhang, Priyanka A., Bharti W. Gawali, Suresh C. Mehrota. *Introduction to EEG- and Speech-Based Emotion Recognition.* Academic Press, 2016.

Abomoelak, B., Prather, R., Pragya, S.U., Pragya, S.C., Mehta, N., Uddin, P., et al. "Cognitive Skills and DNA Methylation Are Correlating in Healthy Novice College Students Practicing Preksha Dhyāna Meditation." *Brain Sciences* 13 (2023): 1214.

Adler, Julius. "The Sensing of Chemicals by Bacteria." *Scientific American*, 243 (1976): 40–47.

Adrados, Francisco Rodríguez. *A History of the Greek Language: From Its Origins to the Present.* Leiden, Netherlands, Brill. 2005.

Agrawal, Pramod Kumar. "A Philosophical Approach toward Electromagnetic Wave Absorption." *Natural Science* 15 (2023): 48–65.

Allen, Richard Hinckley. *Star Names and Their Meanings.* New York: G.E. Stechert, 1899.

"All the Compost Creatures: Levels 1, 2 and 3." *Planet Natural Research Center.* December 8, 2022.

Alonso C., M. Guilarte, M. Vicario, L. Ramos, Z. Ramadan, M. Antolín, et al., "Maladaptive Intestinal Epithelial Responses to Life Stress May Predispose Healthy Women to Gut Mucosal Inflammation." *Gastroenterology* 135 (2008): 163.e1–172.e1.

"Alphabet (Early Greek)." Archaeologies of the Greek Past. Joukowsky Institute for Archaeology. Brown University.

Andriotis, Nikolaos P. *History of the Greek Language.* Servei de Publicacions de la Universitat Autònoma de Barcelona, 2007.

Antonio, Marco Di, Aleks Ponjavic, Antanas Radzevičius, Rohan T. Ranasinghe, Marco Catalano, Xiaoyun Zhang, Jiazhen Shen, et al. "Single-Molecule Visualization of DNA G-Quadruplex Formation in Live Cells." *Nature Chemistry* 12 (2020): 832–37.

Argyropoulos, Eleftherios. *The Alpha Numeric Intelligence of The Hellenic Language.* 2015.

Aversano, Laura. "The Nature of Evil." Unpublished manuscript, 2022.

Bandyopadhyay, Anirban. *Nanobrain: The Making of an Artificial Brainn from a Time Crystal.* Boca Raton: CRC Press, 2020.

Bateson, P. "The Corpse of a Wearisome Debate." *Science* 297 (2002): 2212–13.

Beitman, Bernard. *Meaningful Coincidences: How and Why Synchronicity and Serendipity Happen.* Rochester: Park Street Press, 2022.

Berkeley, University of California. Understanding Science website, "What Has Science Done for You Lately?" Accessed on August 24, 2022.

Besson, Solène, Charles Vragniau, Emilie Vassal-Stermann, Marie Claire Dagher, and Pascal Fender. "The Adenovirus Dodecahedron: Beyond the Platonic Story." *Viruses* 12 (2020): 718.

Blake, William. *Songs of Innocence and Experience.* New York: Start Publishing LLC, 2013.

Braun, Gregory, Dennis Tierney, and Heidrun Schmitzer. "How Rosalind Franklin Discovered the Helical Structure of DNA: Experiments in Diffraction." *The Physics Teacher* 49 (2011): 140–43.

Brophy, Brigid. "Simone de Beauvoir," in her *Don't Never Forget: Collective Views and Reviews,* 285–89. Holt, Rinehart and Winston, New York. 1966.

Brown, William, Amira Baker, and RSF Research Scientists. "The Morphogenic Field is Real and These Scientists Show How to Use It to Understand Nature." *Resonance Science* website. Accessed on August 20, 2022.

Browne, Thomas. *The Garden of Cyrus.* London, United Kingdom, 1658.

Burns, P.D. "Tibetan Singing Bowls for Healing and Meditation." Hubpages website. August 1, 2017.

Caldwell, Shelly. "Our Forests Are Made of Salmon," *North Coast Land Conservancy.* December 9, 2022.

Campbell E.A. "Vibroacoustic Treatment and Self-Care for Managing the Chronic Pain Experience: An Operational Model." Ph.D. thesis, University of Jyväskylä, Yliopisto, Finland, 2019.

Carter, Jake. "Octopus DNA Is Not of This World, Numerous Researchers Conclude," *Anomalien* website. August 22, 2022.

Cayce, Edgar. "Arcturus: A Compilation of Extracts from the Edgar Cayce Readings." Edgar Cayce Foundation. 1971.

Cerciello S., S. Rossi, E. Visonà, K. Corona, F. Oliva. "Clinical Applications of Vibration Therapy in Orthopaedic Practice." *Muscles, Ligaments and Tendons Journal* 6 (2016): 147–56.

Chaix, R., Fagny, M., Cosmin-Tomás, M., Alvarez-López, M., Lemee, L., Regnault, B., Davidson, R.J., Lutz, A., Kaliman, P. "Differential DNA Methylation in Experienced Meditators after an Intensive Day of Mindfulness-based Practice: Implications for Immune-Related Pathways." *Brain, Behavior, & Immunity.* 84 (2020): 36–44.

Charcot, J. M. "Vibratory Therapeutics: The Application of Rapid and Continuous Vibrations to the Treatment of Certain Diseases of the Nervous System." *Journal of Nervous and Mental Disease* 199 (2011): 821–27.

Cochrane, D. J. "Effectiveness of Using Wearable Vibration Therapy to Alleviate Muscle Soreness." *European Journal of Applied Physiology.* 117 (2017): 501–9.

Coxeter, Harold Scott MacDonald. *Regular Polytopes.* 3rd ed. New York: Dover, 1973.

"Croatian Freediver Holds Breath Underwater for Almost 25 Minutes; Breaks Previous Record." *Indian Express*, The Indian Express website, May 18, 2021.

Darwin, Charles. *The Descent of Man, and Selection in Relation to Sex.* 1st ed. London: John Murray, 1871.

Dasanayaka, N. N., N. D. Sirisena, N. Samaranayake, "Association of Meditation with Telomere Dynamics: A Case-control Study in Healthy Adults." *Frontiers in Psychology* 14 (2023): 1222863.

Dayenas, E. et al. "Human Basophil Degranulation Triggered by Very Dilute Antiserum against IgE." *Nature* 333 (1988): 816–18.

Dema, H., A. Paska, K. Kouter, M. Katrašnik, M. Jensterle, A. Janež, A. Oblak, B. Škodlar, J. Bon. "Effects of Mindfulness-Based Therapy on Clinical Symptoms and DNA Methylation in Patients with Polycystic Ovary Syndrome and High Metabolic Risk." *Current Issues in Molecular Biology* 45 (2023): 2717-2737.

Deshmukh, V. "Consciousness, Awareness, and Presence: A Neurobiological Perspective." *International Journal of Yoga* 15 (2022): 144-149.

"Didgeridoo." Duke University Musical Collections. Duke University. May 8, 2024.

"DNA Storage Revolutionary Technology," Sciences Sorbonne Université website. November 20, 2022.

Dossey, Larry. *Healing Words: The Power of Prayer and the Practice of Medicine.* San Francisco: Harper, 1993.

Doughtery, Elizabeth. "Wire Models, Wire. A Brief History of UCSF Chimera." SB Grid Consotium. October 29, 2014.

"Dr. Masaru Emoto and Water Consciousness." The Wellness Enterprise website. March 23, 2017.

Early, Quinn. "Power of the Animals." *Inside Kung Fu*, April 24, 2010.

Feder, Michelle. *Newsela.* Newsela website. "Inventors and Scientists: Dmitri Mendeleev." December 19, 2022.

Feschotte, Cédric and Ellen J. Pritham. "DNA Transposons and the Evolution of Eukaryotic Genomes." *Annual Review of Genetics* 41 (2007): 331–68.

Feuer, Lewis S. "The Dreams of Descartes." Baltimore: The John Hopkins University Press, *American Imago* 20 (1963) 3–26.

Feynman, Richard P. "Plenty of Room at the Bottom." Transcript from the American Physical Society, 1959. Michigan State University website.

Franklin, Rosalind E., and R. G. Gosling. "The Structure of Sodium Thymonucleate Fibres, I. The Influence of Water Content." *Acta Crystallographica* 6 (1953): 673.

Friedlander, John. *Recentering Seth: Teachings from a Multidimensional Entity on Living Gracefully in a World You Create but Do Not Control.* Rochester, Vt.: Bear & Company, 2022.

"Friedrich August Kekulé." Famous Scientists. December 19, 2022.

Galeyev, B. M., and I. L. Vanechkina, "Was Scriabin a Synthesthete?" *Leonardo* 34 (2001): 357–61.

Gao, J., H.K. Leung, B.W.Y. Wu, et al. "The Neurophysiological Correlates of Religious Chanting." *Scientific Reports* 9 (2019): 4262.

Gardner, James. *Encyclopaedia of the Faiths of the World: An Account of All Religions and Religious Sects, Their Doctrines, Rites, Ceremonies, and Customes, Volume 2.* Aryan Books International, 1991.

Gariaev, P., B. Birstein, A. Iarochenko, K.A. Leonova, P. Marcer, U. Kaempf, and G. Tertishy. "The DNA-wave Biocomputer." Fourth International Conference Anticipatory Systems. *International Journal of Computing Anticipatory Systems:* 10 (2001): 290–310.

Gariaev, Peter P., et al., "Materialization of DNA Fragment in Water through Modulated Electromagnetic Irradiation." *DNA Decipher Journal* 4 (2014): 1–56.

Gariaev, Peter P., Peter J. Marcer, Katherine A. Leonova-Gariaeva, Uwe Kaempf, and Valeriy D. Artjukh. "DNA as Basis for Quantum Biocomputer." *DNA Decipher Journal* 1 (2011): 25–46.

Gariaev, P., and M. Pitkänen. "Model for the Findings about Hologram Generating Properties of DNA." *DNA Decipher Journal* 1 (2011): 47–72.

Gilman S. "Joint Position Sense and Vibration Sense: Anatomical Organisation and Assessment." *Journal of Neurology, Neurosurgery, Psychiatry* 73 (2002): 473–477.

Goetz, C. G. "Jean-Martin Charcot and His Vibratory Chair for Parkinson Disease." *Neurology* 73 (2009): 475–78.

Goldman, N., P. Bertone, S. Chen, C. Dessimoz, E. M. LeProust, B. Sipos, and E. Birney. "Toward Practical High-Capacity Low-Maintenance Storage of Digital Information in Synthesised DNA." *Nature* 494 (2013): 77–80.

Gommans, Willemijn M., Sean P. Mullen, and Stefan Maas. "RNA Editing: A Driving Force for Adaptive Evolution?" *Bioessays* 10 (2009): 1137–45.

González-Wippler, Migene. *The Kabbalah and Magic of Angels.* Woodbury, Mn.: Llewellyn Publications, 2013.

Grossinger, Richard. *The Night Sky: Soul and Cosmos.* Berkeley: North Atlantic Books, 2016.

———. *The Return of the Tower of Babel.* Substack, 2022.

———. *Bottoming Out the Universe: Why There Is Something Rather Than Nothing.* Rochester, Vt.: Park Street Press, 2020.

Gurdjieff, G.I. *Views from the Real World.* London: Penguin Books, 1991.

Ha, Melissa, M. Maria, K. Algiers. "Slime Molds," Libre Texts Biology. Yuba College, College of the Redwoods & Ventura College.

Harmony, T. "The Functional Significance of Delta Oscillations in Cognitive Processing." *Frontiers in Integrative Neuroscience* 7 (2013): 83.

Henig, Martin. *Religion in Roman Britain.* London: Routledge, 1984.

Hoelz, André, Angus C. Nairn, and John Kurian. "Crystal Structure of a Tetradecameric Assembly of the Association Domain of Ca²⁺/Calmodulin-Dependent Kinase II." *Molecular Cell* 5 (2003): 1241–51.

Hoffmann A., and H. Gill. "Externally Applied Vibration at 50 Hz Facilitates Dissolution of Blood Clots In-Vitro." *American Journal of Biomedical Science* 4 (2012): 274–84.

Hubbard, Ruth. "Science, Power, Gender: How DNA Became the Book of Life." In *Women, Science, and Technology.* 3rd ed. Hoboken, N.J.: Taylor and Francis, 2013.

Hunt, Catherine. *Your Spark Is Light: The Quantum Mechanics of Human Creation.* Independently published, 2021.

Jia et al., "Long-Term Vegan Meditation Improved Human Gut Microbiota," *Evidence-Based Complementary and Alternative Medicine.* (2020): 9517897.

Jiménez, A., L. Ying, J. Ashwini, L. Galit. "Principles, Mechanisms and Functions of Entrainment in Biological Oscillators." *Interface Focus* 12 (2022): 20210088.

Johnson, Steven. *Emergence: The Connected Lives of Ants, Brains, Cities, and Software.* New York, Scribner Books, 2001.

Kaler, Jim. "Arcturus." University of Illinois Urbana–Champaign. Stars and Constellations. Accessed on May 13, 2024.

Kaliman, Perla. "Epigenetics and Meditation." *Current Opinion in Psychology* 28 (2019): 76–80.

Kantele S., S. Karinkanta, and H. Sievänen. "Effects of Long-Term Whole-Body Vibration Training on Mobility in Patients with Multiple Sclerosis: A Meta-Analysis of Randomized Controlled Trials." *Journal of Neurological Sciences* 358 (2015): 31–37.

Kantor J., L. Kantorová, J. Marečková, D. Peng, and Z. Vilímek. "Potential of Vibroacoustic Therapy in Persons with Cerebral Palsy: An Advanced Narrative Review." *International Journal of Environmental Research and Public Health* 16 (2019): 3940.

Khan, Hazrat Inayat. *The Mysticism of Sound and Music.* Boston: Shambhala Publications, 1991.

Kim H .J., J. H. Kim, Y. J. Song, Y. K. Seo, J. K. Park, and C. W. Kim. "Overexpressed Calponin3 by Subsonic Vibration Induces Neural Differentiation of hUC-MSCs by Regulating the Ionotropic Glutamate Receptor." *Applications of Biochemistry in Biotechnology* 177 (2015): 48–62.

Klimov A., and G. H. Pollack. "Visualization of Charge-Carrier Propagation in Water." *Langmuir.* 23 (2007): 11890–95.

Ko, M.S., Y. J. Sim, D. H. Kim, and H. S. Jeon. "Effects of Three Weeks of Whole-Body Vibration Training on Joint-Position Sense, Balance, and Gait in Children with Cerebral Palsy: A Randomized Controlled Study." *Physiotherapy Canada.* 68 (2016): 99–105.

Koch, Christof. "Activating False Fear Highlights a Memory's Neural Trace." *Scientific American* website. November 20, 2022.

Konkoly, K. R., Kristoffer Appel, Emma Chabani, Delphine Oudiette, and Martin Dresler. "Real-Time Dialogue between Experimenters and Dreamers during REM Sleep." *Current Biology* 31 (2021): 1417–27.

Kopi Luwak Direct website."Arabica and Robusta Varieties and Their Use in the World's Most Expensive Coffees."

Kripalani, S., Pradhan, B., Gilrain, K. "The Potential Positive Epigenetic Effects of Various Mind-body Therapies (MBTs): A Narrative Review." *Journal of Complementary and Integrative Medicine* 19 (2022): 827–32.

Kurek, Michael. *The Sound of Beauty: A Classical Composer on Music in the Spiritual Life.* San Francisco: Ignatius Press, 2019.

Lewis, Paul. *Comic Effects: Interdisciplinary Approaches to Humor in Literature.* Albany: State University of New York Press, 1989.

Liu, Zhenquan, Huina Dong, Yali Cui, Lina Cong, and Dawei Zhang. "Application of Different Types of CRISPR/Cas-Based Systems in Bacteria." *Microbial Cell Factories* 19 (2020): 172.

Lundeberg T., R. Nordemar, and D. Ottoson. "Pain Alleviation by Vibratory Stimulation." *Pain* 20 (1984): 25–44.

Lynch, Jim. Michigan Engineering News, University of Michigan. "Tumors Partially Destroyed with Sound Don't Come Back." Science Daily website, April 18, 2022.

Maeder M., X. Guo, F. Neff, Mathis D. Schneider, and M. M. Gossner. "Temporal and Spatial Dynamics in Soil Acoustics and Their Relation to Soil Animal Diversity." *Plos One* 17 (2022): e0263618.

Mandal, Ananya. "What Is Junk DNA?" *News-Medical* website. November 20, 2022.

Marrin, West. *Water and Nature's Geometry.* The Geometry of Nature Conference (online), 2008.

———. *Universal Water: The Ancient Wisdom and Scientific Theory of Water.* Maui, Hawaii: Inner Ocean Publishing, 2002.

Martin, Jared D., Heather C. Abercrombie, Eva Gilboa-Schechtman, and Paula M. Niedenthal. "Functionally Distinct Smiles Elicit Different Physiological Responses in an Evaluative Context." *Scientific Reports* 8 (2018): 3558.

McMorrow, Erin. *Grounded: A Fierce, Feminine Guide to Connecting with the Soil and Healing from the Ground Up.* Louisville, Colorado: Sounds True, 2021.

Meessen, A. "Virus Destruction by Resonance." *Journal of Modern Physics* 11 (2020): 2011-2052.

Mendioroz, M., Puebla-Guedea, M., Montero-Marin, J. et al. "Telomere Length Correlates with Subtelomeric DNA Methylation in Long-term Mindfulness Practitioners." *Scientific Reports* 10 (2020): 4564.

Meymandi, A. "Music, Medicine, Healing, and the Genome Project." *Psychiatry (Edgmont)* 9 (2009): 43–45.

Morgan J. T., G. R. Fink, and D. P. Bartel. "Excised Linear Introns Regulate Growth in Yeast." *Nature* 565 (2019): 606–11.

Montagnier, Luc, Emilio Del Giudice, Jamal Aïssa, Claude Lavallee, Steven Motschwiller, Antonio Capolupo, Albino Polcari, et al. "Transduction of DNA Information through Water and Electromagnetic Waves." *Electromagnetic Biology and Medicine* 34 (2015): 106–12.

Montagnier, Luc, J. Aissa, E. Del Guidice, C. Lavallee, A. Tedeschi, and G. Vitiello. "DNA Waves and Water." *Journal of Physics: Conference Series* 306 (2011): 0012007.

"Most Proofs of Pythagoras' Theorem." Guinness World Records. May 8, 2024.

"Mysterious Ancient Temples Resonate at the 'Holy Frequency,'" Interesting Engineering website. December 1, 2016.

Navarro, J.F., A. Helmi, K.C. Freeman. "The Extragalactic Origin of the Arcturus Group." *The Astrophysical Journal* 601 (2004): L43.46.

Ningthoujam D. S., N. Singh, and S. Mukherjee. "Possible Roles of Cyclic Meditation in Regulation of the Gut-Brain Axis." *Frontiers in Psychology* 12 (2021): 768031.

Paul, Marla. "Radiant Zinc Fireworks Reveal Quality of Human Egg." *Northwestern Now*, April 26, 2016.

Peacock, Kenneth. "Synesthetic Perception: Alexander's Scriabin's Coloring Hearing." *Music Perception* 2 (1985): 483–506.

Penfield, Wilder. *Mystery of the Mind: A Critical Study of Consciousness and the Human Brain*. Princeton, N.J.: Princeton University Press, 1975.

Plato. E.M. Cope, (trans). *Phaedo*. United Kingdom: University Press, Cambridge, 1875.

———. Lee, Desmond (trans). *Timaeus and Critias*. London: Penguin Books, 1977.

———. Lee, Desmond (trans). *The Republic*. London: Penguin Books Limited, 2007.

Prè D., G. Ceccarelli, L. Visai, L. Benedetti, M. Imbriani, M. G. Cusella De Angelis, and G. Magenes. "High-Frequency Vibration Treatment of Human Bone Marrow Stromal Cells Increases Differentiation toward Bone Tissue." *Bone Marrow Research* 2013 (2013): 803450.

Quaglia, Sofia. "Why Scientists Are Turning Molecules Into Music." *Smithsonian*, online, May 17, 2022.

Radoń, Stanislaw. "The Effect of Meditation Practices on Epigenome Dynamics." *The Unknown Genome, the Latest Research in the Field of Biology and Molecular Diagnostics*. pp. 17–24. Lublin, Tygiel, 2021.

Ren, Z., Q. Lan, Y. Chen, Y. W. J. Chan, G. Mahady, and S. M.-Y Lee. "Low-Magnitude High-Frequency Vibration Decreases Body Weight Gain and Increases Muscle Strength by Enhancing the p38 and AMPK Pathways in db/db Mice." *Diabetes, Metabolic Syndrome and Obesity: Targets and Therapy*. 13 (2020): 979–89.

Sandri M., C. Sandri, A. Gilbert, C. Skurk, E. Calabria, A. Picard, K. Walsh, S. Schiaffino, S. H. Lecker, and A. L. Goldberg. "Foxo Transcription Factors Induce the Atrophy-Related Ubiquitin Ligase Atrogin-1 and Cause Skeletal Muscle Atrophy." *Cell* 117 (2004): 399–412.

Schempp, Walter. "Quantum Holography and Neurocomputer Architectures." *Journal of Mathematical Imaging and Vision* 2 (1992): 279–326.

"Science behind the Wim Hof Method, The." Wim Hof Method website. November 20, 2022.

"Science Historian Cracks 'the Plato Code.'" The University of Manchester website, June 28, 2010.

Sharma P., C. N. Czyz, and A. E. Wulc. "Investigating the Efficacy of Vibration Anesthesia to Reduce Pain from Cosmetic Botulinum Toxin Injections." *Aesthetic Surgery Journal* 31 (2011): 966–71.

Sharp, Gerald. "12 Animals of Xingyiquan." Chi Flow website. December 8, 2022.

Sheldrake, Rupert, Pamela Smart, and Leonidas Avramides. "Automated Tests for Telephone Telepathy Using Mobile Phones." *The Journal of Science and Healing* 11 (2015): 310–19.

Smith, Adam. *An Inquiry into the Nature and Causes of the Wealth of Nations.* Oxford: Clarendon Press, 1979.

Smith, C.W. *Bioelectroydynamics and Biocommunication.* Chapter 3. Singapore: World Scientific Publishing Company, 1994.

Souam, Rabah. "The Schläfli Formula for Polyhedra and Piecewise Smooth Hypersurfaces," *Differential Geometry and its Applications* 20 (2004): 31–45.

Spellman, John. "The Symbolic Significance of the Number Twelve in Ancient India." *The Journal of Asian Studies.* 22 (1962): 79–88.

"Spitz Junior Planetarium." National Museum of American History. Armand Spitz.

Steiner, Rudolph. *Sleep and Dreams.* Great Barrington, Mass.: Steinerbooks, 2003.

Subagia, Nyoman, Gusti Made Widya Sena, and Made Suta. "Comparative Study of Water Before and after Mantra Treatment (Hindu Perspective)." *European Alliance for Innovation.* August 4, 2020.

Tatera, Kelly. The Science Explorer. "5 Dreams That Led to Scientific Breakthroughs and Innovations." *The Science Explorer* website. December 19, 2022.

Taylor, Petroc. "Volume of Data/Information Created, Captured, Copied, and Consumed Worldwide from 2010 to 2020, with Forecasts from 2021 to 2025 (in zettabytes)." Statista.

Torgeson, D. G., Hernandez, R. D. *Rapidly Evolving Genes and Genetic Systems.* Evolutionary signatures in non-coding DNA. Oxford, England: Oxford University Press, 2012.Torgeson, Mark. "Harps for Angels." Mark Torgeson website. December 6, 2022.

Tosenberger, A., N. Besseonov, M. Levin, N. Reinberg, V. Volpert, and N. Morozova. "A Conceptual Model of Morphogenesis and Regeneration." *Acta Biotheretica* 63 (2015): 283–94.

Tyng, Chai M., Hafeez U. Amin, Mohamad N. M. Saad, and Aamir S. Malik. "The Influences of Emotion on Learning and Memory." *Frontiers of Psychology* 8 (2017): 1454.

Underwood, Emily. "Sound of Mom's Voice Boosts Brain Growth in Premature Babies." *Science*, online, February 23, 2015.

U.S. Department of the Interior. "12 Facts About Otters for Sea Otter Awareness Week," U.S. Department of the Interior website. December 9, 2022.

Valéry, Paul, M. Cowley, and J.R. Lawler, (trans). *Collected Works of Paul Valéry, Volume 8: Leonardo, Poe, Mallarme.* Cambridge: Princeton University Press, 1972.

Verdone, L., Caserta, M, Ben-Soussan, T.D., Venditti, S., "On the Road to Resilience: Epigenetic Effects of Meditation." *Vitamins and Hormones* 122 (2023): 339-376.

"Vibratory Therapeutics." *Scientific American.* New York: Springer Nature 1892. p. 265.

Watson, James, Alexander Gann, Jan Witkowski. *The Annotated and Illustrated Double Helix.* New York: Simon & Schuster, 2012.

Watson, James, and Francis Crick. "Molecular Structure of Nucleic Acids: A Structure for Deoxyribose Nucleic Acid." *Nature* 171 (1953): 737–38.

Whiteley, S. J., F. J. Heremans, G. Wolfowicz, D. D. Awschalom, and M. V. Holt. "Correlating Dynamic Strain and Photoluminescence of Solid-State Defects with Stroboscopic X-ray Diffraction Microscopy." *Nature Communications* 10 (2019): no. 3386.

"Why Do Cats Purr?" *Scientific American.* April 3, 2006. Accessed on May 8, 2024.

Wolkstein, Diane and Kramer, Samuel Noah (trans). *Inanna: Queen of Heaven and Earth. Her Stories and Hymns from Sumer.* New York: Harper & Row, 1983.

Wu, D., K. Kenrick, D. Levitin, C. Li, D. Yao. "Bach Is the Father of Harmony: Revealed by a 1/f Fluctuation Analysis across Musical Genres." *Plos One* 11 (2015): e0142431.

Zhang, L., Y. Ren, T. Yang, et al. "Rapid Evolution of Protein Diversity by De Novo Origination in *Oryza*." *Nature Ecology & Evolution* 3 (2019): 679–90.

Zhang, X., M. Bishof, S. L. Bromley, C. V. Kraus, M. S. Safronova, P. Zoller, A. M. Rey, and J. Ye. "Spectroscopic Observation of SU(N)-Symmetric Interactions in Sr Orbital Magnetism." *Science* 345 (2014): 1467–73.

Zheng J. M., and G. H. Pollack. "Long-Range Forces Extending from Polymer-Gel Surfaces." *Physical Review E.* 68 (2003): 031408.

Zhou Y., X. Guan, Z. Zhu, S. Gao, C. Zhang, C. Li, K. Zhou, W. Hou, and H. Yu. "Osteogenic Differentiation of Bone Marrow-Derived Mesenchymal Stromal Cells on Bone-Derived Scaffolds: Effect of Microvibration and Role of Erk1/2 Activation." *European Cells & Materials.* 22 (2011): 12–25.

Index

activation, 28–33, 39–40, 45, 237
Adam and Eve, 180–81
aether, 50, 174–78; *See also* energy/
  energy planes
aging, 105
air, 50, 159–63
allele frequencies, 75
alpha-numerology, ix, 1, 10, 16, 95,
  211, 219, 221, 222
amicable numbers, 18
ancestors, 128–33, 134, 137–39, 141,
  197, 238, 240
angelic realm
  Adam and Eve and, 180
  and ancestral dream ritual, 141
  Arcturus and, 186–88
  communication with, 131, 192–94,
    240
  ferocity and, 195–97
  gematria and, 17
  language of, 197–204
  in meditations, 150, 151, 189–90
animals
  astral bodies of, 128
  communication with, 157, 158, 179

divination via, xi–xii, 156, 159, 178
  in dreams, 178–79
  emotions/intelligence of, 27
  of the five elements, 159–78
  spirit medicine of, 156, 158
anthropomorphism, 27
archetypes, 6, 14, 48, 51, 58, 59, 71
Arcturus, 186–88
Argyropoulous, Eleftherios, 63, 208
astral body/plane, 10, 103, 128–32,
  137–40, 177, 193
Austin, Veda, 84–85, 96
awareness, 56–57

Bach, Johann Sebastian, 98
bacteria, 22, 97, 147, 148
balance, restoring, 64
Bandyopadhyay, Anirban, 50–51
B-DNA, *pl. 7*, 4, 75, 76, 77, 86
bears, 171–72
bees, 161–62
Beitman, Dr. Bernard, 70
Benveniste, Jacques, 84
binary data, 80–81, 181
bison/buffalo, 172–73

Blake, William, x–xi
blood chemistry/circulation, 40, 106
bone health, 106
brain, the
    brainwave oscillations, 90–91, 151
    dodecography (DDG) and, 50–51
    effects of chant on, 29–30
    memory and, 21
    and sound healing, 102, 104, 105,
       115
breath
    as "anima," 156
    and composting grief, 147, 150,
       152–53
    conscious, 7
    cosmic, 114–17
    and DNA, 114, 121–25
    and feeling intention, 112
    healing via, 117–18
    holding your, 127
    intelligence and, 112–13
    and slowing down practice, 46–47
    and sound, 29, 103, 228
    Wim Hof Method (WHM), 119–21
butterflies, 162–63

calm, 136, 147, 164, 225
carbon dioxide, 125
Cayce, Edgar, 187–88
cell membranes, 75–76
change
    and dragons, 176–77
    genetic, 40, 73–74
    inevitability of, 163
    via slowing down, 45
    temporary, 110
    vibrational, 11

chant, 29
chaos, 76, 82
child, inner, 31–32
coincidence, meaningful, 70
cold exposure, 120
collectivity, 58, 59–60
color/sound system, 90–91, 115–17
commodification, 36, 155
complexity, 34–37, 45–47
composting grief, 146–51
conception, 115, 116
consciousness
    and the astral plane, 128
    and the breath, 114, 115
    and chant, 29
    and the dodecahedron, 56–57
    evolution of, 12, 74, 81
    and grief, 143, 145
    and memory, 20, 22
    and quality of life, 36
    of the soul, 115
    states of, 129–30
    of water, 84, 85
cosmic breath, 114–17
cosmic communication, 15, 191–92
courage, 79, 167, 173, 193, 198, 200,
    203
coyotes, 170–71
creation
    channeling, 180
    and diversity/mutuality, 60
    encrypting/decrypting, 16, 62
    life as conferred by, 6
    of man in God's image, 18
    and natural sigils, xi
    at a personal scale, 10
    and sound, xv, 100

Crick, Francis, 2, 4, 13, 72, 86
CRISPR, xiv
crows, 160–61
cubes, 50–52, 87–89, 225, 226, 228
cymatics, 95

data storage, 80–81
death, 28, 132
deer, 173–74
defective genes, 73–74
devolution, 34, 36–37, 45–47
divination, x, 16–17, 156, 159, 178
divine, the, 54, 191–92, 222–24, 227
DNA
  angelic communication and, 194
  and archetype recreation, 71
  and the breath, 114, 117–18, 127
  chemical structure of, *pl. 1–3*, 13, 38
  and consciousness, 12
  discovery of double helix, 2–5,
    13–15
  and the dodecahedron, 53–54
  experience as coded in, 24
  gematria of, 1, 182, 241
  genes and, 5–6, 16
  and Hellenistic legacy, 66
  how to honor, 2
  influencing your, 6, 10, 241
  memory and, 19, 21–22
  mitochondrial DNA, 144–45
  noncoding DNA, 72–74
  palindromes in, 15–16
  and RTS (rapid technology storage),
    69–70
  sound and, xiii–xiv, 11, 30, 94, 234
  splicing, 37
  as telepathic, 77–78

water and, 57, 86, 89
  See also 12-stranded DNA
dodecahedron
  and the brain, 50–51
  at the cellular level, 55–56
  and consciousness, 56–57
  as evolutionary model, 2
  and the five elements, 50
  and icosahedron, 87–89, 91, 95
  illustration of a, *pl. 4*
  in meditation practices, 209
  twelves in culture/nature, 52–56
dogs/wolves, 173
double helix structure, 2–5, 13–15, 76
dragonflies, 176
dragons, 176–77
dreaming, 129, 133–42, 178–79

eagles/hawks, 175–76
Earth/Gaia, 50, 146–48, 171–74, 203,
  205
economics, 35
ego, 59
elders, 178, 238
electromagnetic fields, 77, 78, 191
elements, 50, 159–78, 185
emotion
  in animals, 27
  benefits of, 202
  and brain waves, 104
  calming your, 136
  and cymatics, 95
  emotional memory, 22
  and healing, 28, 43, 46, 101, 103–4
  meditation and, 149, 151
empathy, 22, 26–27, 43
empowerment, 7, 21, 142

energy/energy planes
   astral plane, 10, 103, 128–32, 137,
      138, 177, 193
   changes at the level of, 11, 239
   divine energy field, 222–24
   and immortality, 28
   and medicine, 93, 95
   morphogenetic fields, 71–72
   and the physical, 10
   and primal nadis, 14
   scalar energy, 78
   and traditionary sciences, 63–64
entities, individual, 37
entrainment, 29–30, 104
entropy, 76
epigenetic regulation, 39–40
equality, 26–27
eroticism, 82
evolution
   and accepting mortality, 28
   creative, 98
   of defective genes, 73–74
   devolving for, 34, 36, 58
   and DNA activation, 10, 39–40,
      234–35
   and the dodecahedron, 2
   frequency of, 11, 229–30
   genetic and cosmic, 82
   and the God code, 6, 7
   as healing, 43, 110, 142, 236, 238
   lack of mindful, 1
   and miracles, 239
   and morphogenetic fields,
      71–72
   rapid, 146
   symbols to aid personal, 10
EZ (exclusion zone), 85

fear, 45, 46, 82, 142, 173
feeling, 112
Fibonacci sequence, 88, 117, 182,
   185–86, 223, 226
fire, 50, 167–71
fireflies, 171
fish, 165–66
five elements, 50, 159–78, 185
fluidity, 15, 45
Forms, 49–51
fourth dimension, 88–89, 234–35
foxes, 168–69
fractals, 70–71, 72
Franklin, Roslind, 2–5, 14, 86
freedom, 113
frequencies. See vibration
friends/brothers, 232/233
frogs, 165

geese, 163
gematria
   of Adam and Eve, 180–81
   of DNA, 1, 241
   of dreams, 134
   efficacy of, 16–17, 235
   encrypting/decrypting, 62
   of Forms, 49–51
   of God and immortality, 18–19,
      182–83
   historic origins of, 68
   isopsephia as equivalent, 73
   meditations, 209–20
   of new life, 117–18
   of the New Sound Frequencies,
      225–34
   and Plato's ideal world, 31
   of rapid, 71

generational healing, 20, 23, 24, 145, 240
genes
  breath to enhance, 127
  changing expression of, 40
  damage to, 46
  DNA and, 5–6, 114
  editing, 39–40
  evolution of, 73–74, 82
  grief's effect on, 143, 144, 145
  healing our genome, 144
  memory and, 21, 22, 73
  original form of, 48
  transposons in the genome, 73
  viruses and biological memory, 37
God, 18, 182–83, 189
God code, 6, 7, 183
golden ratio, 87–88, 185, 226, 230
Gosling, Raymond, 2, 4
gratitude, 59, 150, 151, 160
gravitational constant, 44
Greek language, 1, 65–68, 222
grief, 23, 46, 104, 143–53, 238, 239, 240
Grossinger, Richard, xv, 81, 130, 156, 170
Guido d'Arezzo, 183, 222
gut-brain axis (GBA), 149

harmony, 186, 223, 228–29, 231, 234
healing
  and the breath, 112–13, 117–18
  and composting grief, 146–51
  of DNA, 6, 7
  via dreams, 137
  as evolution, 110, 142, 238
  generational, 20, 23, 24, 240

  and medicine, 155
  via memory, 25–26, 28
  as nonlinear, 150
  organized vibrational medicine, 95–96
  and prayer, 118
  sound healing, xiii, xv, 85, 99–108, 136, 138
heart, 22, 153, 241
Hellenic language, 1, 65–68, 222
hero, missing, 58–61
higher self, 22, 23
hiraeth, 29
Hof, Wim, 119
Holy Spirit, 213–14, 233–34
hope, 43, 44
humans, 21, 128–29, 156, 194, 239, 241
hummingbirds, 159–60
hydroglyphs, 84–85
hyperventilation, 120, 125

icosahedron, 50–52, 54–55, 57, 87–89, 225
identity, 66, 81–82
imagination, 30–32
immortality, 18, 23–25
immune system, 148–49, 169
Indigenous nations, 155
inflammatory response, 120
information
  brain and retrieval of, 21
  and consciousness, 13
  conveyed in dreams, 134, 138
  and entropy, 76
  molecular-level, 73
  in scalar energy, 78

self-organized, 34–35
shared with the ground, 151
trials to extract deep, 42
inner child, 31–32
instinct, 72, 200
intelligence
"brainless," 237–38
and the breath, 112
cosmological, 221, 235
of DNA and water, 86
human, 241
and necessity, 57–58, 59
shared in dreams, 134–35
intention, 92, 96–97, 101, 112, 135, 141
interconnection
of all living forms, 26–28
experience of, 47
via mantra meditation, 30
and the missing hero, 58–59
and morphogenesis, 188
sacred mindframe of, 240
and universal consciousness, 57
in water, 83, 85
interventions, 201
isopsephia, 73, 134, 189, 211, 212, 215, 216, 218

Jenny, Dr. Hans, 95
Jung, Carl, 48, 59, 70

Kennedy, Dr. Jay, 64, 65
Kopi Luwak, 36

Langridge, Robert, 53
language
angelic, 197–204

ESL (English as a second language), xii
and gematria, 17
Greek, 1, 65–68, 222
palindromes, 16
laughter, 201–2
letting go, 98, 99, 103, 104, 138, 142, 152, 165–66, 167
lies, 44–45
life
breath and flow of, 113
complex self-organization of, 34–35
as interconnected, 26–28, 165–66
as miraculous, 239
natural healing in, 110
new, 115, 116, 117–18, 188, 228–29
of passion and purpose, 20
lions, 167–68
love, 41, 42, 44–45, 127, 132, 194–95
lucid dreaming, 129, 133–34, 135–37, 138, 141, 142

macrocosm/microcosm, xi, 2, 5, 13
manifestation, 25
mantras, 29–30, 96–97, 123
marmots, 174
matter, 12, 36, 49, 191, 193, 240
meaning, 11
meditations
to activate 12-stranded DNA, 39–40
angelic water meditation, 189–90
and animal/human exchange, 158
for astral body engagement, 139–40
and breath/vibration, 46
and composting grief, 146–51
gematria meditations, 209–20

and immunity, 148–49
inner child meditation, 30–32
and lucid dreaming, 136
slowing down practice, 46–47
and theta rhythm, 102
memory
  and the ancestral plane, 137
  and the brain, 21
  consciousness and, 20–21
  and DNA, 19, 78, 114
  editing your, 37–43
  emotional, 22
  of genes, 73
  grief and, 145, 238
  and healing, 20, 23, 25–26, 132
  of hope, 44
  immortality and, 24–25
  learning as remembering, 49
  in morphogenetic fields, 71–72
  practices to activate, 28–33
  of the soul, 33, 42
  and theta rhythm, 102
  of water, 84
mentality, 232
microbiome, 146, 147, 148, 149, 150, 151
Mind, the, 24
mindfulness practices, 40, 112, 120
miracles, 238
missing hero, 58–61
mitochondrial DNA, 144–45
Montagnier, Luc, 77
morphogenetic fields, 69, 70–75, 76, 188
mortality, 24, 25, 28, 112
mRNA (messenger RNA), 38, 39, 44
mugwort, 141

musculoskeletal system, 107–8, 125–26
mutuality, 59–60
myths, 132

nature, xi, 10, 13, 52, 93, 99, 146–47
necessity, 57–58, 59
nervous system, 126
neurological health, 106–7
New Sound Frequencies, 221, 223, 225–34
nine, power of, 224–25
noncoding DNA, 72–74, 145
nondualism, 56, 209
nucleic palindromes, 16
numbers/numerology
  amicable numbers, 18
  and divination, 17, 219–20
  and gematria, 62–63
  of nine, 224–25
  numbers as vibrations, 180–81
  power of, 6
  wisdom in numbers, 63
  See also gematria

octahedron, 50–52, 224, 225
octave, 225
oracles. *See* divination
organized vibrational medicine, 95–96
otters, 166–67
out-of-body experiences, 129, 138
oxidative stress, 145
oxygen, 125–26

pain, 107, 110, 132
palindromes, ix, 15–16, 18, 69–70
paradigms, 49

pentagrams/pentagons, 55, 96, 185, 226, 229, 230

perfection, 49

Plato
five solids of, 17, 50, 51, 56, 184–85
gematria of, 184–85
and the ideal world, 31
on intelligence and necessity, 57–58
musically encrypted writings of, 64–65
Platonic Forms, 49

Pollack, Gerald, 85

polytopes, 88

prayer, healing power of, 118

proteins, 38, 55–56

purity, 228–29

purpose, 28

Pythagoreans, x, 7, 17, 50, 64, 208, 225, 230

quantum imprinting, 77–78

quantum symmetry, 89–90

regression, 198–201

resonance, 103, 156

retracing, 58

RNA (ribonucleic acid), 38

RTS (rapid technology storage), 70–81

scalar energy, 78

science
author's career in, 74–75, 79
of gematria, 63
and the Greek language, 67
of healing prayer, 118–19
limited understanding of, 14–15, 240
vs. spiritual practice, 40–41

self-organization, 34–35

sigils, x, xii

skin, 30

slime mold, 13, 34, 237–38

slowing down, 45, 46–47

smiling, 124, 126, 159–60

snake medicine, 169

soil, 146–48

Solfeggio frequencies, 183, 222, 223, 224

solids, Platonic, 17, 50, 51, 56, 88, 184–85

souls, 28, 33, 115, 116, 133

sound
and the astral body, 131
and breathwork, 121–25
and composting grief, 149–50, 151–53
and DNA, xiv, 78
dodecahedron frequencies, 222–24
healing, xiii, xv, 85, 99–108, 136, 138, 239
intentionality about, 10
and lucid dreaming, 136
of mantras, 29–30
and memory activation, 28–33
music of Plato's books, 64
New Sound Frequencies, 225–35
and skin, 30
Solfeggio frequencies, 183, 222
sound-to-color system, 90–91, 115–17
sound waves, 227
and traditionary sciences, 63
and water, pl. 9–13, 91, 94, 95–98, 221
of your voice, 152

space, 45, 47, 88–89
spheres, harmony/music of, 64, 98
spin symmetry, 89–91
spirals, 57, 77, 81, 231
spirit, divine, 155
spiritual practices, 40–41
Steiner, Rudolf, 10, 130, 221
storage, 69–70, 80–81
storytelling, 132
strands, DNA, 69–70
stress, 46, 144, 145
suffering, 43, 130–31, 132
swans, 166
synergy, 19
systems, 34–37, 77

tears, 202–3
technology, 69–70, 75–78, 91, 239
telepathic exchanges, 129, 194
Tetragrammaton, 189
tetrahedron, 50–52, 87, 224, 232
theta rhythm, 102
thought
    as adaptive, 45
    and cymatics, 95
    empowering conscious, 21
    and genetic state, 146
    on immortality, 23
    and mindfulness, 34
    as unspoken words, 29
time, 10, 11, 30, 45, 160–61
traditionary sciences, 63
trances, 129
tranquility, 225
transposons, 73
trauma, 24, 144, 145
tRNA (transfer RNA), 38, 39

trust, 45, 142, 202, 203
truth, 44–45, 79
turtles, 164
12-stranded DNA
    activating, 11, 39–40, 41, 114,
        121–25, 238
    as the God code, 7
    illustration of, pl. 5–6
    inner understanding of, 47
    language of explaining, 67–68
    means for exploring, 5, 6
    meditations on, 39–40, 189–90,
        209–20
    as the missing hero, 59
    palindromes in, 16
    as path to immortality, 19
    sound frequencies of, 222–24,
        225–34
    sound-spinning molecule of, 91
    and water, 14

unconscious, the
    and DNA, 12, 22
    and dreaming, 129, 133
    immortality as, 24, 28
    meditation to invoke, 139
    sound as activating, 28
    and truth, 44

vagus nerve, 126
vibration
    angelic, 192–93, 197–98
    and cymatics, 95
    and generational healing, 23
    healing via, 99–108
    of love, 41, 44
    and mantras, 29

New Sound Frequencies, 221, 223, 225–34
organized vibrational medicine, 95–96
particle movement via, 94
raising a soul's, 28, 39–40
space needed for, 46
and traditionary sciences, 63–64
vibrational change, 11
See also sound
viruses, 37, 91, 143–44
visualization, 30–32, 123, 139–40
vultures, 177–78

waking, 128–29
water
    animals of, 163–67
    consciousness of, 84, 85, 93
    and DNA structure, 14, 86, 89
    and the dodecahedron, 57
    and the five elements, 50
    and the hero's journey, 60–61
    hydroglyphs in, 84–85

and the icosahedron, 54, 55, 57, 87–89, 91, 95, 117
imprinting, 92–93
meditation on, 189–90
memory of, 84
and organized vibrational medicine, 96
phases of, 85, 87
sound's effect on, *pl. 9–13*, 91, 94, 95–98
vibrations in/on, 30
Watson, James, 2, 3, 4, 13, 86
wave genetics, 78
wellness, 39–40, 101, 111
wholeness, 111, 221
Wilkins, Maurice, 2, 3, 4, 14
will, the, 2, 24, 113
Wim Hof Method (WHM), 119–21
wisdom, 63, 134
work, 161–62
world, the, x, 7, 31, 49

zinc, 115, 116, 117

About the Author

Ruslana Remennikova is a writer and sound healer originally from Kyiv, Ukraine. Her early interest in nature, science, and chemistry led her to pursue a bachelor's degree in biochemistry and biology at Virginia Commonwealth University. During her undergraduate years, she studied zebrafish embryology and worked as a research assistant in a microbiology and immunology laboratory at Virginia Commonwealth University School of Medicine. Remennikova then went on to earn her master's degree in chemistry from the University of North Carolina at Wilmington.

In 2010 Remennikova began a ten-year tenure at the pharmaceutical contract research organization Pharmaceutical Product Development (PPD Inc.), where, as a senior scientist and, later, an associate research scientist, she worked with norovirus, dengue virus, and coronavirus. PPD later became part of the Fortune 100–ranked company Thermo Fisher Scientific.

In 2016, Remennikova experienced a life-changing call to seek a more meaningful life and a career outside of the science industry, which resulted in her becoming an all-world Ironman triathlete while opening a health-based food truck, Pulp RVA, and later a coffee shop, Pulp on Lakeside. In 2019 Pulp RVA was selected as a contestant by the Food Network for a feature on the top fifty food trucks in the United States.

She currently resides in Richmond, Virginia, where she founded and operates Songbird Science, specializing in research and sound medicine to nurture well-being and cultivate peaceful, forward-thinking learning communities.